Psychological Theory and Teaching Profession

About the Author

L.B. Tripathi, has retired from Gorakhpur University as Professor and Head, Department of Psychology where he started his career as Assistant Professor in 1957. He held various administrative positions at Gorakhpur University including Dean, Arts Faculty (1985-88) and Pro Vice Chancellor and acting Vice Chancellor (1990-91). He was a Commonwealth Scholar in Canada (1960-63) and a Visiting Professor at Psychology Department and Centre for Behavioural and Cognitive Science (CBCS), Allahabad University (1999-2003). In addition to 40 papers and 4 book chapters, Dr. Tripathi has authored *Psychological Consequences of Prolonged Deprivation* (with G. Misra), *Calibration of Verbal Material in Hindi and Psychological Outcomes Associated with Severe Iodine Deficiency* (with B.D. Tiwari), *Adhunik Prayogik Manovigyan, Adhunik Samaj Manovigyan, and Manovaigyanik Anusandhan Padhati*. He has edited *Manovaigyanik Samkhyiki*, and *Daihik Manovigyan*. He has translated *Psychology of Learning* by Hulls, Deese and Egeth in Hindi. He can be contacted at HIG-32, Pritam Nagar, Allahabad-211011, Tel: 0532-2230500.

About the Editor

Girishwar Misra, Ph.D., is Professor and former Head, Department of Psychology, and Dean Faculty of Arts, University of Delhi. He has been Fulbright Fellow at Swarthmore College and the University of Michigan. During his academic career spanning over three decades, he has previously taught at the Universities of Bhopal, Allahabad and Gorakhpur. The Government of Madhya Pradesh conferred upon him the Radhakrishnan and Dr. Hari Singh Gaur Awards. Professor Misra has authored or edited *Psychological Consequences of Prolonged Deprivation, Deprivation: Its Social Roots and Psychological Consequences, Applied Social Psychology in India, Perspectives on Indigenous Psychology, Psychological Perspectives on Stress and Health, Psychology of Poverty and Social Disadvantage*, and *Contributions to Psychology in India*. He has been President and Convener of the National Academy of Psychology, India. He is currently editing its journal *Psychological Studies. He can be contacted at* misragirishwar@gmail.com.

Psychological Theory and Teaching Profession

Selected Essays by

Prof. L.B. Tripathi

Editor

Girishwar Misra

CONCEPT PUBLISHING COMPANY, NEW DELHI-110059

Cataloging in Publication Data—DK
Courtesy: D.K. Agencies (P) Ltd. <docinfo@dkagencies.com>

Tripathi, L. B. (Lal Bachan), 1935-
Psychological theory and teaching profession :selected essays by Prof. L.B. Tripathi / editor, Girishwar Misra.
p. cm.
Includes bibliographical references.
Includes index.
ISBN 8180692906

1. Psychology. 2. College teaching. 3. College teaching—India. I. Misra, Girishwar, 1951- II. Title.

DDC 150.954 22

ISBN 81-8069-290-6

First Published 2006

Published and Printed by
Ashok Kumar Mittal
Concept Publishing Company,
A/15-16, Commercial Block, Mohan Garden,
New Delhi-110059 (INDIA)

Phones : 25351460, 25351794
Fax : 091-11-25357103
Email : publishing@conceptpub.com

Contents

Introduction

Girishwar Misra

Psychology as a modern discipline has imbibed the spirit of science for its disciplinary deliberations in the academia. As such it aspires for unequivocal, dependable and objective approach towards understanding the observable psychological reality. A large majority in the college of psychology believes so, and is quite hopeful that such an approach shall help in achieving the goal of certainty in the pursuit of knowledge by removing the ignorance existing on account of non-scientific and faith oriented exercises. In such a frame of mind science happens to be the *sine qua non* of the evolution and growth of knowledge in the modern world. It is assumed to allow the pursuit of the goals of understanding, control and prediction in a deterministic and mechanistic paradigm in a value neutral mode. A strong commitment to this position has been the hallmark of the academic and professional concerns of Professor L.B. Tripathi.

Professor Tripathi appreciates the value of science and maintains vigorously its spirit in his academic approach and personal conduct. He believes in gaining knowledge through empirical methods and techniques rooted in objectively verifiable domains which set the contours of reality and knowledge. His remarks in the foreword to Professor K.D. Broota's *Experimental Design in Behavioral Research* (1989, Delhi: Wiley Eastern) articulate his methodological position:

> *"Experimental method is the mainstay of social science research in general and psychological research in particular, mainly because it is considered to be the most dependable way of determining causal relationships between the antecedent and the consequent variables."* (p. v)

In the true scientific spirit Professor Tripathi considers powerful theorisation capable of explaining increasingly greater domain of an area of study as the prime concern of scientific pursuit. He views science as an exercise in building theories that have potentials of high degree of generalizability. Its not a mere coincidence that he designed, introduced and taught a compulsory course on psychological theory at Master's level at Gorakhpur University and opted for Theoretical Psychology as one of the fields of specialisation for the Department when the UGC recogzised it as a Department of Special Assistance (DSA).

Another concern of Professor Tripathi has been the study of the dynamics of teaching profession. He chose this profession, enjoyed it and set a model to be emulated by others. He has examined many aspects of the lives of teachers in the institutions of higher learning. He has been reflecting on the changing Indian scenario in this profession. His reflections, rooted in his observations and contemplative analyses have insights to reorganize higher education. Throughout his career he strived to maintain the dignity of the teaching community and shared his concerns on various fronts.

Keeping the core of Professor Tripathi's twin concerns and emphases that characterize his academic pursuit, it was decided to bring out an anthology of his select writings. As one can notice the deliberations in the chapters of this volume revolve primarily on two themes, i.e. developments in psychological theory and teaching profession. They encompass a wide range of issues pertaining to the developments in mainstream psychology, culture, deprivation, parenting and the changing scenario in the professional culture in the community of teachers. The articles on teaching profession included in this volume present a vivid picture of the changing scenario and address the problems of values, accountability, conduct and aspirations of teachers in Indian Universities. The analysis reveals the story of decline in the academic standards, increasing political interference, and overall weakening of the structure of higher education in the country. Professor Tripathi also comes out with certain concrete suggestions to improve the situation and bring back the lost glory of the temples of learning.

The original sources of the papers included in this volume are as follows.

PART I
PSYCHOLOGICAL THEORY

Chapter 1: Information theory in contemporary psychology. This was originally presented at a seminar on psychological theory organized by Late Professor Raj Narain, at Lucknow University, Lucknow. It was subsequently published by Professor M.C. Joshi in *Indian Psychological Review*, 1966, *2*, 109-118.

Chapter 2: Current theoretical developments in experimental psychology. This was published in *Indian Journal of Current Psychological Research*, 1991; *6*, 1-12.

Chapter 3: Culture as a psychological construct. This was presented at the Seminar on Culture and Psychology organized by Professor Girishwar Misra at Delhi University in 1997. It was published by Professor R.C. Tripathi in *Psychology and Developing Societies*, 2001, *13*, 129-140.

Chapter 4: Some reflections on contemporary psychology. This was an invited article originally published *in Psychological Studies*, 2000, *45*, 14-18.

Chapter 5: Knowledge and faith in contemporary psychology. This was an invited article published in *Psychological Studies*, 2003, *48*, 5-16.

Chapter 6: Some methodological problems of deprivation studies. This was a paper originally presented at the Seminar on Deprivational Studies organized by Late Professor Durganand Sinha at the University of Allahabad. It was subsequently published in D. Sinha, R.C. Tripathi and G. Misra (Eds.) *Deprivation, its social roots and psychological consequences.* (pp. 49-71), New Delhi: Concept.

Chapter 7: Some erroneous beliefs of parents about their children. It was a keynote address at the Professor N.S. Chauhan Memorial Seminar on Parental Behaviour and Child development organized by Praachi Psycho-cultural Research Association and Institute of Home Science, B.R. Ambedker University, Agra and subsequently published in *Praachi Journal of Psycho-cultural Dimensions*, 1999, *15*, 91-94.

Chapter 8: A Theoretical Perspective on Man-Environment Relationship. It was originally presented at a Seminar on Environmental Psychology held at Bhopal University in 1988.

Chapter 9: Psychology and Social Problems. Inaugural address given at the Annual Convention of National Academy of Psychology at K.S. Saket P.G. College, Faizabad in March 2005.

PART II
TEACHING PROFESSION

Chapter 10: Some motivational aspects of university teachers. It was a key note address at a Seminar on Motivation in Educational Context organized by Professor Girishwar Misra at Bhopal University, Bhopal.

Chapter 11: On reward and punishment in Indian universities. *University News, 1999, 37(10),* 1-5.

Chapter 12: Extra remunerative assignments in the universities. *University News, 2004, 42(30), 1-4.*

Chapter 13: Roles and responsibilities of university teachers. *Zenith,* A journal published by A.P.S. University, Rewa, 9-13.

Chapter 14: Indian universities underlying assumptions and consequences. *Amity Business School Review, 2000, 1,* 6-12.

Chapter 15: My concept of a university. *Zenith,* A journal published by A.P. S. University, Rewa, 12-21.

In order to situate these contributions in perspective it seems appropriate to learn about the person behind them as well as the socio-cultural context in which the themes emerged and took concrete shape. Professor Tripathi was born on 15th September, 1935 in a village named Dharahara in Balia district of Uttar Pradesh, a region famous for achievements in the Indian scholarly tradition as well as contributions to the national freedom struggle and Indian polity. His father Late Sri Jagannath Tripathi was a farmer by profession. After early education at a Sanskrit Pathshala in a nearby village, the traditional centre for learning, he received higher secondary education in Azamgarh. He completed higher education at

Banaras Hindu University, Varanasi under scholars like Professors B.L. Atreya and S.S. Jalota. After doing Master's in psychology with the distinction of being first class first in 1955 he started a career in the professional world at D.J. College at Barot near Meerut in western Uttar Pradesh. While working at Barot he was selected as an Assistant Professor at Gorakhpur University in 1957 by Sri Bhairava Nath Jha, an educationist of great repute. He along with Dr. Madan Mohan Sinha and Sri S.B.L. Bhardwaj started the Department of Psychology at Gorakhpur University.

The turning point in his academic career came when he left for Canada as a Commonwealth Scholar. He was trained as a psychologist at McGill University, Montreal, under the pioneers of modern psychology including Donald O. Hebb and Dalbir Bindra. He earned doctorate in Human Cognitive Psychology based on his research on concept formation under Professor Bindra in 1963. After completing his doctoral studies he joined Gorakhpur University in 1963 and moved to Jodhpur University as Reader and Head of Psychology Department for a year. He returned to Gorakhpur as Reader in 1964 and became Professor and Head in 1968.

As a university teacher, Professor Tripathi's goal was to train young minds with an indefatigable zeal. He committed himself to the mission of teaching and research in the field of Psychology. As a hard task master he nurtured the students and colleagues in a selfless manner. His presence in the Department helped to create an atmosphere of sincerity, genuineness and dedication to research and academic work. After serving the university for 39 years in various capacities including Dean, Pro Vice Chancellor and acting Vice Chancellor he retired in 1996. On the invitation of Professor R.C. Tripathi he joined the Centre for Advanced Studies (CAS) in Psychology at Allahabad University from 1999 to 2001 as a Visiting Professor. Subsequently, Professor Janak Pandey requested him to be associated with the newly founded UGC Centre for Behavioural and Cognitive Sciences (CBCS) as a Visiting Professor. He continued in that capacity from October 2002.

Professor Tripathi initiated his students, colleagues and peers in methods, theorization and different domains of research and equipped them to pursue research and apply the resulting knowledge for addressing various social problems. Commitment to scientific reasoning coupled with critical and comprehensive understanding comprised his motto. At the University of Gorakhpur he guided numerous students for doctoral work. Twenty-two students earned doctorate under his direct supervision. They worked in various areas of basic psychological processes including *time perception* (Professor Sidha Nath Rai), *person perception* (Professor Bhaskar Deo Mishra), *prolonged deprivation* (Professor Girishwar Misra, Dr. Ashtanand Tiwari, Dr. Sushil Kumar Tiwari), *social motivation* (Professor Narendra Kumar Mani Tripathi, Dr. Rashmi Gupta, Dr. Pushpa Sharma), *time perspective* (Professor Adesh Agarwal), *learning and memory* (Dr. Sheela Singh, Dr. Ramji Lal Srivastava, Dr. Veera Srivastava, Professor Anupam Nath Tripathi), *religiosity* (Professor Achal Nandini Srivastava), *group polarization* (Dr. Rajendra Tiwari), *dependence proneness* (Professor Banarasi Das Tiwari, Dr. Ramesh Mani Tripathi) and *child rearing and socialization* (Dr. Renu Singh, Dr. Madhulika Srivastava). It is gratifying that most of these students are now senior faculty and some of them have extended the domains of research and have earned a place in the academic world. In addition to his own students he has been active in promoting research and encouraging scholars to work in different areas in various capacities.

Professor Tripathi undertook a number of research projects including *Calibration of Materials in Hindi for Research Purposes* (ICSSR), *Psychological Consequences of Prolonged Deprivation* (with Girishwar Misra) (ICSSR), *Learning Disability in a Deprived Society* (with Achal Nandini Srivastava) (NCERT), and *Psychological Outcomes Associated with Severe Iodine Deficiency* (With Banarasi Das Tiwari) (ICMR).

His presence at the UGC as member of subject panel and UGC National Lecturer in Psychology was widely appreciated. He was part of the Editorial Committee of the First ICSSR Survey of Research in Psychology under the editorship of Late Professor Shiva K. Mitra. Subsequently he wrote a review chapter on *Higher Mental Processes* in the third ICSSR Survey.

He has been contributing to ensure quality of research as a member on the editorial boards of many journals including

Indian Journal of Psychology, Indian Psychological Review, Indian Psychologist, Indian Journal of Current Psychological Research, Personality Study and Group Behavior and *Psychological Studies.*

In his professional life he was associated with Indian Science Congress Association. He was also the Executive of Indian Psychological Association (IPA) but did not engage in the institutional politics and refrained himself from such activities. However, he took the initiative to organize the psychological community in Uttar Pradesh. He is the founder President of U.P. Psychological Association (UPPA) and continued in this capacity for five terms. Looking at the political and personal diversions in the professional organizations, psychologists gathered at Bhopal University in 1988 decided to form the National Academy of Psychology (NAOP) India. On that occasion Professor Tripathi was unanimously nominated as the convener of advisory committee for drafting the constitution of the new organization.

As a teacher he felt the need to provide authentic material for students and committed himself to this goal. As Hindi became the medium of instruction in the Universities of the Hindi belt, he along with his colleagues took the initiative to bring out a series of standard text books in Hindi. His book *Modern Experimental Psychology* has been printed for 10 editions. He also wrote and edited text books on *Social Psychology, Physiological Psychology, Research Methods* and *Psychological Statistics*. He translated the Hulse, Deese and Egeth's *Psychology of Learning* in Hindi which was published by Tata Mcgraw-Hill (India).

Professor Tripathi started his university career when the challenge was to develop an identity for the newly emerging discipline of psychology and putting it on a sound footing in the university community. Realizing this responsibility and foreseeing its wider implications he worked hard to establish this discipline in the world of higher learning. He became instrumental in starting psychology courses in numerous institutions and revising and updating the courses. He invested his energy in preparing the ground for the take off of this newly growing discipline. He established a model centre of higher learning in psychology at Gorakhpur. Under his dynamic leadership, this Department was elevated by the UGC to the status of Department of Special Assistance (DSA).

In addition to teaching and research, Professor Tripathi has been and continues to be a motivating force. Scholars often seek advice and guidance in academic matters from him. Professor Tripathi happily responds to such roles. Thus when late Prof. (Mrs.) Anima Sen was writing the monograph on *Attention and Distraction*, (1983, New Delhi: Sterling), she asked Professor Tripathi to write the foreword of the book. Professor Tripathi happily obliged her by providing a lucid and in-depth analysis of cognitive psychology in general, and attention in particular. His foreword is a very precise and conceptually rich piece. To have a feel of his style of exposition I would like to quote from this foreword. He observes:

> *"Gradual realization of the failure of the behaviouristic paradigm, and irresistible upsurge made by the cognitive psychology provided the necessary impetus and ideas for revival of some older mentalistic concepts with more readily operationalizable definitions. Great strides made in the field of information processing technologies and communications proved extremely fertile for yielding new models for understanding the functions of human brain. ... By the end of the 60s psychology had gone a full circle from the point of abandoning mentalistic concepts to reaccepting them* "(pp. x-xi).

When Professor Tripathi was asked to review research in Higher Mental Processes in the third ICSSR survey he had to organize the chapter and develop a conceptual scheme. As such the term "higher mental processes" (against lower cognitive processes!) does not occur in psychological vocabulary, so Professor Tripathi (1988, p. 289) had to develop a notion that could be defended in the scientific discourse. He proposed that

> *"higher mental processes are those psychological activities in which mediational processes go beyond the direct constraints of sensory input and motor output, are predominantly determined by neurophysiological processes of the brain, and are explicable in terms of either reductionistic neurophysiological processes or special constructs."*

[Higher mental processes. In Janak Pandey (ed.) (1988)

Psychology in India: The state of the art (Vol. 1, pp. 289-328) New Delhi: Sage.]

It is because of clarity in conceptualization and expository style that Professor Tripathi happens to be one of the most sought after resource persons at National seminars and refresher courses in psychology and UGC orientation programmes for the university teachers. He is regarded as an undisputed scholar in the field of human experimental psychology and people come to him to clarify their concepts. His concern for viewing themes in the historical perspective and relating them to the broader theoretical context makes him the central figure in the academic debates and discourses that he joins.

A champion of panhuman psychology rooted in positivist tradition of thought, Professor Tripathi's scholarship has continued to enrich the perspectives of students and researchers. Informed by the vision that foundational psychological processes are the common building blocks of psychological discourse he finds neuropsychological and cognitive processes more basic. He is convinced that these processes are potent enough to explain the socio-cultural processes. With an open but critical mind he comes out with a powerful scrutiny of many of the assumptions, beliefs, values and preferences held by the majority or masses. He believes in parsimony and finds many concepts not only theoretically superfluous but scientifically untenable.

As a person, Professor Tripathi commands unparalleled respect from colleagues and students. He reciprocates others but without compromising his own identity. Those who have visited him cannot forget his hospitality, concern, care and warmth. He draws attention from academicians and leaves of an indelible impression on the audience. His academic preferences and scientific commitments, however, do not come in the way of personal relationships. He is fondly remembered as a dependable friend and guide by the people who came in his contact, irrespective of social status, caste or creed. Living in the complex culture like India where collectivism and individualism both are supposed to coexist, he recognizes his cultural roots but does not allow them to overpower or disrupt his scientific commitment. They do not come in the way of appreciating and respecting the values and cultural practices held by others.

During a long inning as a leader in the discipline of psychology at the national scene spanning over a period of four decades he has been a constant source of inspiration and a guiding force for a large number of colleagues and students. As a Guru, Professor Tripathi has nurtured several generations of psychologists. Engaged in the endeavour of institution building and pursuing the Dharma of a mentor, he invested his entire energy in teaching and helping people learn to do scientific psychology. He has not only witnessed the discipline grow from very close quarters but has also participated in shaping it.

Keeping the concerns of Professor Tripathi in view it was considered appropriate to bring out a selection of his writings and make them available to a wider readership. The present volume is an outcome of such an effort. I take this opportunity to express my gratitude to Professor Tripathi who allowed me to organize this project. Sri Ashok Mittal of Concept Publishing Company readily agreed to publish the manuscript. He deserves thanks for supporting this endeavour. I am confident that the readers would find this volume useful in understanding theoretical and professional issues and feel encouraged to relate to the social world that they inhabit.

Part I

Psychological Theory

1

Information Theory in Contemporary Psychology

Inherent shortcomings of both S-R and cognitive theories in providing a scientifically satisfactory explanation of behavioural organization and sequence on the one hand and tremendous success of man-made devices in stimulating complex psychological processes on the other hand led a number of psychologists to believe that operational principles of behaviour can be better understood and analyzed in terms of servo-mechanism and information. The hypothesis that "the nervous system is a communication network, the brain functions as a computer with extensive facilities for information storage, and the psychological processes are information-handling operations" (Attneave, 1959) originated from the writings of Norbert Weiner (1948) and Shannon (1949) and has supportive evidence to its credit from the investigations of George Miller (1951, 1953, 1956), Pollack (1952, 1953) and Bourne and Restle (1959). The impact of this hypothesis may be described as three fold on the contemporary psychology. First, the information theory as a formal branch of Cybernetics, provided a fruitful quantitative technique for measuring various facets of psychological processes in a more precise way than the traditional approaches. The information theory, as a matter of fact, provided a model for designing and analyzing psychological experiments. Second, this approach led a number of psychologists to formulate new theories of processes such as perception, signal detection, concept identification and strategies in categorization behaviour etc. Included in this impact is an attempt by Miller, Gallanter and Pribram (1960) at construction of full-fledged system of behaviour in their book

Plans and the Structures of Behavior which has received wide recognition from psychologists. In this book the authors have liken a human being with a computer which is a self-programming and programme-operating system. They have also attempted at an informational analysis of complex processes such as instinct, memory, intention and language behaviour. The third impact—most fashionable but not equally commendable—is the use of informational style of thinking and speculation accompanied with informational vocabulary. The current psychological writings are replete with terms like redundancy, input-output, feedback, decision-making tactics and strategies, coding-decoding etc. The fashion of analogous vocabulary is so well practised by the readers that "well worn phrases like perceptual organization, disorganizing effects of emotion, knowledge of results, stereotyped behaviour, reorganization of problem material etc. leap off the pages" (Miller, 1953). Apart from these direct impacts one more important though indirect is the current trend in physiological psychology. A number of researchers have started looking at the human nervous system as a network of communication operating on the principles as envisaged in the information theory. This influence can be summarily observed in a book entitled *The Machinery of the Brain* by Wooldridge (1963).

This paper presents a brief account of the in-roads of information theory in psychology. With this aim in view, first, a short characterization of information theory is presented. While doing so the basic concepts and formulae are omitted for lack of space and time and also for the reason that excellent accounts of the same are available elsewhere. Second, an account of some experiments is given which have been done in the framework of the theory. Third, summary accounts of some psychological information theories are presented. Finally, the prospects of information theory in psychology and computer-communication models for psychological experiments are discussed.

Information Theory

The information theory as a branch of cybernetics is a formal as opposed to a substantive-empirical-theory. The concept of information is quantitative and not intuitive, and is used as a measure of 'message' in communicational technology to establish equivalence amongst messages and to limit the degree of association or correlation between them. In this sense the theory is an extension of the correlation theory. Quantitatively it is concerned with stochastic processes which have to be ergodic or in which the laws of probability characterizing the processes have to remain constant throughout the sequence. The theory as such does not assume substantive-predictive characteristics.

In its application to the study of behaviour it reflects an attitude or opinion and analyzes behaviour as a sequence of random selection of sensory input-output (or stimulus-response if you will). This random selection follows probability laws. Hence, behavioural sequence is a stochastic process. The theory in this framework provides an experimental model for conceptualizing behavioural sequences and organizations as sensory inputs-outputs with a variety of relevancy and redundancy involved in them, and for designing ways of attacks on psychological problems. In theorization the information-theoretic approach looks for the principles by which subject selects and codes the stimuli, and achieves effectiveness in processing these coded inputs. Thus psychological information theory is different from information theory of cybernetics.

Informational Experiments

A sizable number of experiments have been reported which demonstrate that traditional as well as newer problems of experimental psychology can be more precisely and accurately tackled in the framework of information model in which a responding organism is conceptualized as a communication channel with definable capacity for handling, storing and modifying information input. In this model the information input (call it sensory stimulation if you will) contains determinable amount of redundancy and irrelevancy. In the process of its encoding and channelization the channel creates noise that results in modified information output. Within this broad framework informational experiments in psychology have been conducted. Let us examine some of them.

Channel Capacity Span of attention or apprehension or immediate memory is an old problem. Experiments on channel capacity are analogous, though more precise, to such problems. Informational inquiry in this connection has tackled two questions, *i.e.* the maximum rate at which man can transmit information and second, the maximum amount of information that can be conveyed to an observer by means of a single stimulus dimension or by several dimensions in controlled combinations. A familiar theorem of information theory specifies that information can be sent at the channel capacity rate (C) with as small amount of error or equivocation as desired by proper encoding. However, if rate is more than C (say C+R) there will necessarily be an equivocation equal to or greater than RI, the excess. With a view to determine human C, varied amounts or bits of information have been put into adults at varied rates and more or less consistent Cs with small variability have been obtained in a variety of experimental conditions.

Pollack (1952) using the method of absolute judgment, required subjects to identify tones which assumed values in logarithmic steps from 1000 cps to 8000 cps and had to assign numerals to them. The input information varied from 1 bit to 3.5 bits and the transmitted information under the optimal condition was 2.5 bits at the most. With increase in input information the output increased linearly upto 2 bits and then bent off to an asymptote at 2.5 bits which may be considered as the C. Pollack investigated this C in variety of ways by controlling the spacing of tones and judgmental conditions. The results were always more or less than same, 2.5 bits or 6 alternatives.

The C for transmission of visual positions has been investigated by Hake and Garner (1951). In their studies subjects had to identify positions of a pointer between two end-marks labelled as 0 and 100. The experiment was conducted under two conditions. In one the response alternatives were unlimited within the given range and pointer positions limited, while in the other the response alternatives were limited to the pointer positions. The result was that subjects could transmit 3.5 bits of information or 9 alternatives. Coonan (see Miller, 1956) repeated the Hake-Garner experiment and obtained 3.2 bits for very short exposures and 3.9 for long exposures.

Halse and Chapnis (1954) have found that C for absolute judgment of colours ranges from 3.2 to 4.08 bits. Eriksen (1954) found C of 2.2 bits for sizes, for hue 3.1 and for brightness 2.3 bits. Beebe-Centre *et al.* (1959) investigated the C for taste intensities with concentrations of salt solutions ranging from 3 to 34.7 NaCl per 100 cc tap water in equal subjective steps and found it to be 1.9 bits or 4 alternatives.

Miller (1956) has summarized the findings of Pollack group at the Air Force Operational Applications Laboratory. They have found Cs for areas 2.6 and 2.7 bits for length of lines, 2.6 to 3.00 bits for directions or angle of inclination, 2.8 and 3.3 for short exposures and long exposures respectively. The lowest C was found to be 1.6 bits for curvature if the length of the arc was constant but it increased to 2.2 bits when instead of length are was kept constant.

This survey leads us to believe that the C of man is limited and does not show wide range of variability. Whether owing to the limitations of learning or built-in-limitations of the nervous system man as an information handling system operates on a relatively narrow channel. This obvious conclusion might make you sceptic. It is common experience that we recognize any of thousands of words, faces or objects and the C appears to be quite infinite. Miller (1956) opines that this is so due to multidimensionality of objects and words we recognize. Data are available to show that greater the dimensionality of an object, greater becomes the C for absolute judgment of their variates. Klemmer and Frick (1953) found the C of 4.6 bits for judging the positions of a dot in a square—a tridimensional situation. Beebe-Centre *et al.* (1955) found the C of 2.3 bits for both the sweetness and saltiness of concentrations with various amounts of salt and sucrose. For judgment of loudness and pitch of tones Pollack (1953) fond the C of 3.1 bits. Pollack and Fricks (1954) managed to combine six dimensions in judged tones and showed the C to be 7.2 bits. Thus our doubt is dispelled. It seems to be confirmed that the C increased as a function of the number of dimensions in an object's recognition and location. However, this function is neither multiplicative nor additive. As a matter of fact the exact mathematical relationship has not yet been worked out. Miller (1956) suggests that the human C is limited, but man as an information-processing system manages to break or stretch this judgmental bottleneck by organizing the stimulus-input-simultaneously into various dimensions and successively into a sequence of chunks of information.

Concept-learning in the Informational framework. Usually in a concept formation paradigm subject is confronted with a set of stimuli (one at a time) of which each stimulus consists of specified variate of a variety of dimensions or continua; and he is required to identify each stimulus as belonging to either-or a number specified categories. When the subject identifies the set of stimuli to a criterion measure, he is said to have learned the concept. Implicit is in this paradigm is the idea that conceptualization is basically a categorization task. It is intuitively clear that categorization involves partitioning of stimuli, a process upon which the entire methodology of information theory rests. Because of this commonality psychologists did not spare anytime in attacking the problems of concept formation through information methods and analyses.

Pioneering efforts in this direction came from Hovland and Weiss (1952, 1953) who worked on the role of positive and negative instances of concepts as information transmitters, though without measuring information in bits. Bourne and associates took the lead and converted the area of concept learning in a rich land for information theoretic approach. They have investigated the effects of informational variables such as redundancy, complexity, irrelevancy, feedback of information, compactness of patterns etc. Let us review a couple of these experiments.

It is intuitively clear that irrelevant information input would create equivocation in the output till the irrelevant bits are sorted out, differently coded and adapted to. Information-theoretic orientation would suggest that greater would be the sorting time resulting in increased number of errors in concept learning tasks. Archer, Bourne and Brown (1955) experimented on concept-identification with two bits of relevant and 1, 2 and

3 bits of irrelevant information containing set of stimuli or instances. They found that the performance was a positive exponential function of the amount of irrelevant information. This finding has been confirmed by Bulgarella and Archer (1962). The inhibitive role of irrelevant information has been laid bare by experiments which reported noisy instances with identification of recurrent regularities.

Most of the sensory input which we channelize in our everyday life contains great amount of redundant information bits some of which being relevant facilitate our tasks and being irrelevant make our tasks more difficult though easier than non-redundant irrelevancy does. Redundancy, in the words of Kochen (in Hunt, 1963) refers to non-information transmitting attributes of stimuli. We have, for example, 75 per cent redundancy in the English language (Shanon, 1951). Bourne and Heygood (1959) demonstrated that redundancy in the relevant attributes of concept instances facilitated the identification task, because it increased the types of cues and thereby the channel capacity. In the same study it was also shown that redundant irrelevancy interfered with the conceptualization task.

Apart from these comparatively more obvious information experiments on concept identification, a number of experiments have been reported on the role of informational variables in concept identification which are of less obvious nature (see Bourne, 1964 and Hunt, 1962).

Other Experiments—So far I have presented a very sketchy review of experimental work on channel capacity and concept formation in the information theoretic framework. Within this framework a number of other psychological problems have also been experimented upon with exciting results. As a matter of fact, information model has become a tool to tackle problems in the most desirable quantitative manner, which once were considered unexperimentable. Attneave (1959) has pointed out that one of the great fascinations of information theory is that it has provided a methodology for quantifying Gestaltists' subjective concepts of structure, organization or patterning once considered to be unquantifiable. Hochberg and McAllister (1953) reported, after experimenting with Kopferman's Cubes, that "the probability of a perceptual response is an inverse function of the amount of information required to define that pattern". Weistein and Fitts (1954) used histogram like patterns varying in informational content and reported that identification time increased with complexity. Bricker (1955) found similar results in a paired associate learning experiment in which stimulus redundancy was added by increasing extra elements. Attneave (1955) in a series of three experiments on recognition, delayed memory etc., found that figural goodness is equivalent to redundancy.

In a number of experiments Averbach and Sperling have studied short-term storage of information in vision. The guessing game technique of information theory introduced by Shannon has been used to study rates at which information is transmitted by individuals. Some researchers have shown linear relationship between per stimulus information presentation time. Klemmer and Frick (1953) found that the rate of continuous information transmission increased with number of alternative stimuli.

Information theoretic model has provided most promising tools for study of psychophysical problems. It is well known that despite innumerable studies we never arrived at a stable measure of absolute limen for any sense modality and were almost convinced that RL is a variable phenomenon. However, some researchers have determined or developed techniques to determine stable level of thresholds for individuals. They have used the information model and the concepts of sensory input and noise to their fullest extent. The studies too are many to be easily reviewed. As a matter of fact that area of information theoretic investigations is so broad that they cannot be reviewed here even sketchily. Therefore, now we should proceed to some theoretical formulation in informational framework.

Theories

Psychological information theories are numerous and have emerged from different sources. At the very outset it must be pointed out that most of such theories are in formative stage and therefore, cannot be properly evaluated. Some of the theories have deviated from the formal and mathematical bases of the information theory owing to a variety of reasons; but all of them start positing theorems or stating propositions from identical theoretical foundations. They all regard, though implicitly, that human organism is an information processing computer. The computer receives information (sensory input) and transmits it (response output). All inputs contain relevant and/or irrelevant information with varying degree of redundancy. During the encoding process the organism creates interferences (noise) resulting in loss or dissipation of input information (analogous to entropy). By virtue of repeated information encoding and sequential decision making the organism more or less like a digital computer develops schema or more economical ways of coding further information inputs. This is what learning is in the informational framework. Since the guiding idea behind such theorization is that operational principles of computer and human organism are alike, human organism is conceived as storing information and schema in the manner of a computer.

Now we may review some of such theories.

Bruner's Informational Approach to Cognition—Bruner and his co-authors of "*A Study of Thinking*" (1956) toyed with the informational ideas and terminology in analysis, understanding and explanations of thinking and conceptualization problems. The concept learning or attainment was defined as a series of decision making processes, nay, they urged that "virtually all cognitive activity involves and is dependent on categorization process" (p. 246). However, the spirit of the book was more akin to the sign-gestalt theory than to modern information theoretic model, though Bruner freely used the informational phraseology without taking recourse to their quantifying principles. It is only in his later work (Bruner 1957a) that he seriously started theorizing in the formal informational framework.

Proposing a theory of perception in general and perceptual readiness in specific, Bruner (1957a) states that perception "depends upon the construction of a set of organized categories in terms of which stimulus input may be sorted, given identity, and given more elaborated connotative meaning". While analyzing perception in terms of seven propositions, he posits that perception is a decision making process for categorization in which discriminatory cues are used. By cue utilization he implies going from cue to an inference of identity and assignment to "either-or" or "one of the many" categories. Categories refer to set of specifications or rules for equivalence. For an organism all categories are not always equally easily available, rather they vary in terms of accessibility or readiness with which stimulus inputs with given attributes will be coded in a category. Veridicality in perception consists of objective appropriateness of a category. Encoded input will be veridical to the degree to which the accessibility of categorizing systems reflects, the likelihood of the occurrence of the events that the person will encounter.

For such perceptual categorization Bruner posits four mediating mechanisms the first of which is grouping and integration the physiological analogue of which are best explained through the concepts of cell-assembly and phase sequence (Hebb, 1949). The next mechanism is access ordering, a device for making the categories accessible for assignment of the sensory input to one of them. The third mechanism is match-mismatch signalling. It is through this mechanism that a particular categorization is examined for accuracy. The final one is the gating mechanism which monitors or filters information at the peripheral level and send it to the higher centres.

In another paper Bruner (1957 b) has extended this theory to a condition of perceptual categorization where the given input is not sufficient for a successful categorization. In other words, the input is incomplete or gapy and still the organism fills in the necessary information bits and makes a successful categorization. According to Bruner this type of perceptual gap filling or extrapolation is done through code-systems which are learned through repeated processing on input information and thereby acquiring schema, as Bartlett would put it. Importance of such scheme or code systems has been recognized by Hebb (1949) and Attneave (1959). A coding system is defined as "as a set of contingently related non-specific categories". Acquisition of such coding systems depends on set, need state, degree of mastery and its relationship with general coding and diversity of training.

Bourne and Restle's Theory of Concept Learning—Bourne and Restle (1959) presented a theory of concept learning which they called a mathematical theory. The principles of this theory with respect to learning were derivations of stimulus sampling and cue conditioning. The phraseology which they used and the techniques Bourne later employed were those of information theory. Stimuli are specified by binary dimensions the values of which may become relevant or irrelevant and/or redundant. Relevant values are supposed to transmit sufficient information for correct decision or categorization. Reinforcement or feedback of information produces conditioning of relevant cues and adoption of irrelevant cues. Probabilities of correct responses on trials are posited as mathematical functions of numbers of relevant-irrelevant cues.

The theory as Hunt (1962) points out is limited to the analysis of conjunctive concepts and the model assumes path independence which is of course assumed by all stimulus sampling models. Some experiments have demonstrated the inadequacy of this model and the data are now more impressive in favour of hypothesis testing model as compared to the cue-conditioning model. *S-R and Information Theory: Theoretical Integration by Berlyne:* Berlyne (1957) having found impressive evidence in favour of information approach attempted an integration of S-R. theory with information theory.

He points out that information theory has proved useful in those problems where subject is confronted with a binary decision situation or decisional conflict. The study of conflict in the framework of behaviour theory, thus, becomes the point of contact between the two theories. Varying degrees of conflict creates varying degrees of uncertainty and *vice versa*. One of the major functions of information theory is to measure the amount of such uncertainties. Correlated with degrees of conflict are emotional disturbance, reaction time, drive, curiosity, stimulus complexity and positive attraction of risk taking. Study of these S-R problems may prove more useful in the framework of information theory.

Miller's Plan Theory of Behaviour—It is well-known that George Miller was one of the first psychologists who brought the formal information theory to the realm of psychological problems. Miller, Gallanter and Pribram (1960) presented a programatic theory of behaviour in general. Its concepts are derived from both communication and computer theory. The main emphasis is on cognition and sequential-hierarchical organization of behaviour. It assumes that "behaviours are guided by plans which are hierarchical processes in the organisms that can control the order in which a sequence of operations is to be performed" (Miller *et al.* 1960, p. 16). It is clear that a program is to a computer, as a plan is to an organism. The molar aspect of behaviour execution of a plan is in sequential order. Image is collected or built-in store of information about the world or itself in the organism. The unit of behaviour for analytical purposes is proposed a TOTE hypothesis which is advanced as a substitute for S-R or Sign-Significance unit. TOTE is an ecronym for Test-Operate-Test-Exit. Each TOTE is a decision unit which when compounded gives rise to binary decisions.

Within this basic framework the authors define intention as the uncompleted task whose execution has just begun; habits and skills as plans which through learning becomes involuntary from voluntary. Such plans are usually integrated through experience and relinquished when necessary.

A TOTE unit is abstracted as information transmission from one part of the unit to the other; planning as information handling at the higher levels; and execution transmission at the muscular level. Most of the plans are learned and learning is not establishment of associations between S and R or S and S but is development or acquisition of information processing techniques.

Epilogue

By way of summing up it may be stated that the information theory is rapidly gaining ground in contemporary psychology; but it is still too early to present a coherent and complete picture of its contributions. It is undoubtable that this theory has broken new grounds in psychological experimentation, provided new tools of data analysis, and more meaningful explanatory concepts. It is rather encouraging in this respect that physiological researches have provided supportive evidence for approach. Wooldridge (1963) states that... "computers and the brain do not display superficial similarities in some of their operational characteristics. Instead they are the mechanisms of the same kind, in the sense that they obtain their similar results by essentially similar means". However, still more work has to be done before we could built up a realistic hope that this approach will resolve the mysteries of human behaviour.

References

Arches, E.J., Boourne, Lyle, E., & Prowne, F.C. (1955). Concept identification as a function of irrelevant information and instructions *Journal of Experimental Psychology, 49*, 153-164.

Attneave, Fred (1959). *Applications of Information Theory to Psychology*, New York: Holt-Dryden.

Beebe-Centre, J.G., Rogers, M.S. & O'connel, D.N. (1955). Transmission of information about sucrose and saline solutions through the sense of taste. *Journal of Psychology, 39*, 157-160.

Berlyne, D.E. (1957). Uncertainty and conflict: A point of contact between information theory and behavior theory concepts. *Psychological Review, 64*, 329-339.

Bourne, Lyle E., Jr. (1959). Informational Variables in Concept Formation BRL *Report No. 36*, University of Colorodo, U.S.A.

Bourn, L.E., & Heygood, R.C. (1959). The role of stimulus redundancy in concept identification. *Journal of Experimental Psychology, 58*, 232-38.

Bourne, L.E., & Restle, F. (1959). Mathematical theory of concept identification. *Psychological Review, 66*, 278-297 (a).

Bricker, P.D. (1959). The identification of redundant stimulus patterns. *Journal of Experimental Psychology, 49*, 73-81.

Bruner, Jerome, S. (1957a). On perceptual readiness. *Psychological Review, 64*, 123-52.

Bruner, Jerome, S. (1957b). On going beyond the information given, in *Contemporary Approaches in Cognition*. Cambridge Mass: Harvard University Press.

Bruner, J.S., Goodnow, J.J., & Austin, G.A. (1956). *A Study of Thinking*. New York: Wiley

Bulgarella, Rosaria G., & Archer, J.E. (1962). Concept identification of auditory stimuli as a function of amount of relevant and irrelevant information. *Journal of Experimental Psychology, 63*, 254-257

Eriksen, C.W. (1954). Multidimensional stimulus differences and accuracy of discrimination, U.S.A.F. Wadc. Tech. Reg. No. 54-163.

Hake, H.W., & Garner, V.R. (1951). The effect of presenting various members of discrete steps on scale reading accuracy. *Journal of Experimental Psychology, 42,* 358-366.

Halse, R.M., & Chapnis, A. (1954). Chromaticity, confusion contours in a complex viewing situation, *Journal of Optical Society of America, 44,* 442-454.

Hebb, D.O. (1947). *The Orgination of Behaviour,* New York: Wiley.

Hochberg, J.E., & McAllister, E.A. (1953) A quantitative approach to figural goodness. *Journal of Experimental Psychology, 46,* 361-364.

Hovland, Carl J. (1952). A communication analysis of concept learning. *Psychology Review, 59,* 461-472.

Hovland, Carl I., & Weiss, W.J. (1953). Transmission of information concerning concepts through positive and negative instance. *Journal of Experimental Psychology, 54,* 175-182.

Hunt, Earl B. (1962). *Concept Learning: An Information Processing Problem.* New York: Wiley.

Klemmer, E.T., & Frick, F.C. (1953). Assimilation of information from dot and matrices patterns. *Journal of Experimental Psychology, 45,* 15-19.

Miller, George (1951). *Language and Communication,* New York: McGraw Hill.

Miller, George (1953). What is information measurement? *American Psychologist, 8,* 1.

Miller, George (1956). The magical number seven, plus or minus two: some limits on our capacity for information processing. *Psychological Review, 63,* 2.

Miller G. Gallanter, E., & Pribram, Karl H. (1960) *Plans and the Structures of Behavior*. New York: Holt.

Newman, E.B. (1951) Computational methods useful in analyzing series of binary data. *American Journal of Psychology, 64,* 252-262.

Pollack, D. (1952). The assimilation of requentially encoded information. *American Journal of Psychology, 66,* 421-435.

Pollack, L. (1953). The information of elementary auditory displays. II. *Journal of Accoustic Society of America, 25,* 765-769. (a).

Pollack, J., & Fricks, L. (1954). Information of elementary multidimension auditory displays. *Journal of Accoustic Society of America, 26,* 155-158.

Shannon, E. (1949). Communication in the presence of noise. *Procceedings of Institute of Radio Engineers, 37,* 10-21.

Shannon, E. (1951). Prediction and entropy of printed English. *Bell Systems Technical Journal, 30,* 50-64.

Weiner, N. (1948). *Cybermetics*. New York: Willey.

Weiner, N. (1950) *The Human Use of Human Beings*. Boston: Houghton Mifflin.

Weistein, M., & Fitts, P.M. (1954). A quantitative study of stimulus complexity in visual pattern discrimination. *American Psychologist,* 49 (abstract).

Wooldridge, Dean E. (1963). *The Machinery of the Brain*. New York: McGraw Hill.

2

Foundations of Recent Developments in Experimental Psychology

The present theme assumes vital importance because the contemporary psychology in general, and experimental psychology in particular, is in the process of a complete transformation. Voluminous experimental data have been generated necessitating reformulation of metatheoretical assumptions that underly the theoretical propositions, constructs and concepts, as well as paradiagmatic thinking and data collection procedures. This paper shall endeavour to delineate the historical background of those events, discoveries of facts and the theoretical shortcomings that created the compelling need for re-orienting theorization and conceptionalization mental activities.

For 40 years, that is, from 1920 to 1960 approximately, the analysis of human behaviour or mental functions had been carried out under the predominant influence of a learning paradigm the foundations of which were laid down and solidified by Thorndike, Pavlov, Hull, Skinner, Guthrie, Spence, Miller, Hebb and a host of the other psychologists. Under the influence of the Darwinian biology the gap between man and animal was minimized, and was converted from qualitative to quantitative. The metatheoretical assumptions underlying this paradigm were simple and in consonance with the then intellectual zeitgeist. From amoeba to man were assumed to be under the control of external environment, technically called stimulus. Behaviours were assumed to be reactions or responses to these stimuli. In experimental research living organisms were given experimental treatment and were called subjects. It is self-evident that the very semantics of the paradigm illustrates the passive role assigned to the living organisms. The appetitive and aversive conditions and reinforcements were assumed to determine the nature and extent of S-R associations which constituted the basis of behaviours that experimental psychology was developed up to 1960.

It is true that a number of psychologists such as Tolman and numerous Gestalt psychologists resisted the spread of these assumptions and endeavoured to develop alternative paradigms and methodological prescriptions and prescriptions created by the associationistic behaviourists. The feeble voice of protest raised by them became integrated into the main stream of experimental psychology. The elegance, parsimony and scientific aura of the learning paradigm so impressed the dissenters, that MacCorqudale and Meehl (1954) tried to translate the sign-gestalt theory of Tolman in the vocabularly and format of the behaviourists. Its unavoidable consequence was that the entire domain of experimental psychology became overwhelmingly behaviouristic. Topics smacking of mentalism were banished from the realm of psychology. Perception and conception were translated as discrimination learning. Thinking and reasoning became peripheral. Attention was no more considered a necessary part of psychological activities.

During this process experiments were carried out to verify and extend the theoretical propositions put forward by learning psychologists. In numerous areas the obtained data failed to conform to the predictions and expectations of behaviouristic theories. In many areas the data led to such diverse interpretations that theoretical contradictions and confusions made many young psychologists lost in it. Let us examine some of the areas where interpretational problems forced psychologists to do rethinking about theoretical assumptions.

There were two principles of learning which were universally accepted. The first, as Seligman (1970) called it, was the *associability principle* and the second was the *contiguity principle*. According to the associability principle the nature of CS in the Pavlovian and the discriminative stimulus or the discriminative stimulus or the discriminanda in the instrumental learning is irrelevant and matter of indifference. It is assumed that any stimulus can be associated with any learnable response. According to the second principle, for learning to occur the interval between CS and US and analogously between the instrumental response and the occurrence of reinforcement has necessarily to be a few seconds (less than 5 sec.) at the most. Psychologists have successfully generated experimental data that prove the empirical validity of these two principles. In this regard a series of experimental studies reported by Garcia and his associates (1955, 1966 & 1972) and Rozin and Kalat (1971) can be cited. These investigators have unequivocally demonstrated that a rat develops conditioned aversion for the once preferred saccharin solution even though there is delay of several hours between the taste of the sweetened drink (CS) and the toxic effect of X-irradiation (negative reinforcement). In another set of experiments these investigators have demonstrated the genuineness of the stimulus relevance phenomenon which forcefully refutes the empirical validity of the associability principles. Shettleworth (1972) has also demonstrated that certain stimuli alone and not other stimuli get associated or conditioned with certain responses. For example, in the Garcia study sweet as well as bright-noisy water was injected while the rats were being X-irradiated. After a few hours rats developed toxic effects which resulted in learned aversion for sweet water, but not for the bright-noisy water. The findings of long delay learning and the stimulus relevance phenomena were contrary to the well established assumptions of learning paradigm, and necessitated change in the theoretical assumptions.

Let us scan some other areas in experimental psychology where experimental findings have gone beyond the interpretational scope of the main behaviouristic theory. Most of us are aware that the interference account of forgetting is basically anchored in the S-R theory, whether it is conceived as the independence hypothesis of McGeoch (1932), or as the unlearning and interference model of Melton and Irwin (1940) or the response set theory of forgetting initiated by Newton and Wickens (1956) and elaborated by Postman, Stark and Fraser (1968). In these versions of the theory the association and the response aspects only were invoked to explain both proactive and retroactive forgetting. Martin (1971) on the basis of his findings questioned the predictive adequacy of these theories and proposed an alternative theory apparently within the same metatheoretic framework. The theory has come to be known as 'encoding variability theory of learning and forgetting.' The most interesting aspect of this theory is that a distinction is made between the nominal and the functional stimulus. A moments reflection brings to notice that this nominal stimulus is the objective physical stimulus as presented by the experimenter, and the functional stimulus is the one which has been encoded by the subject, distinction as an indirect acceptance of the role of cognitive or perceptual variables in forgetting. It is indeed indicative of the explanatory inadequacy of the theories of forgetting anchored in the learning paradigm.

There is another interesting trend which has damaged the validity of the interference theory beyond repair. Tulving and his associates attacked the details of the interference theory and contrary to the predictions of the interference theory produced experimental data to prove contextual biasing, more effective recall than recognition, and lack of retroactive forgetting in presence of correct retrieval cues. The data led him to propound a new theory of forgetting known as the cue-dependant forgetting theory. This theory, in brief, states that forgetting is not due to interference but due to non-availability of appropriate cues at the time of retrieval. Anderson and Bower (1973) have presented a new associationistic model of memory. None of these theories are able to explain all the data available on memory. All of these mini-theories have been experimentally demonstrated to have explanatory shortcomings.

Now let us take up another case from the domain of learning paradigm, that is, the concept of reinforcement. We know that for Thorndike reinforcement was creating a satisfying state of affairs, but for Hull it was drive stimuli reducing. On the contrary Sheffield showed that reinforcement induces drive and lies in occurrence of vigorous consummatory activity. Olds and Milner (1954), Olds (1958), and Bower and Miller (1958) have shown that reinforcement both positive and negative, are brain stimulations. Apart from these reductionistic explanations based on solid empirical data, there are constructive interpretations also. For instance Premack (1959, 1971) has developed an interesting theory. He states that 'given two responses, one of which is more probable than the other, an opportunity to perform the more probable task will provide for reinforcement of the less probable response'. This point of view has also been subjected to criticism and related new theories have emerged such as Timberlake and Allison's (1974) response deprivation hypothesis. The pertinent question is what is reinforcement? Naturally, the learning paradigm does not have a definitive answer.

The confusion about reinforcement becomes truly confounded when we look at some other aspects of reinforcement. Menzel (1973) and Bandura (1965, 1971) have advocated the concept of vicarious reinforcement with a view of explaining the occurrence of imitative learning in the incentive conditions of the learning paradigm. Vicarious reinforcement is interpreted as reinforcement experienced indirectly through observing reinforcement being provided to another organism. One can easily realize that the limitations of the learning paradigm forced these psychologists to traverse in the cognitive paradigm in the cognitive realm though they avoided using the cognitive terms. Subsequently after having done more studies, Bandura (1978) was forced to move farther and farther away from the learning constructs and in 1978 he came to a conclusion that self-reward does not strengthen the response that preceded it. Now Mischel, Kanfer and Bandura as well as a host of others emphasize the informational property of the event described as 'reinforcing', and have abandoned the reinforcement concept as utilized earlier.

A few words more about the infirmity inherent in the reinforcement functions. Reinforcement, irrespective of the diversity and contradictions involved in its theoretical interpretations, was assumed to strengthen the response that precedes it. Some psychologists such as Deci (1975) and Lepper and Green (1978) have evolved the concept of intrinsic motivation and intrinsic reward. They have also shown that extrinsic reward on performance of intrinsically interesting activities causes a loss of interest in the task. Hence, the obvious conclusion is that under certain conditions extrinsic reward weakens the response. On the other hand, Rosenfield, Folger and Adelman (1980) have shown circumstances in which extrinsic reinforcement enhances motivation especially when extrinsic reward is perceived as reflecting personal competence. The experimental findings go beyond the reinforcement interpretation in the learning paradigm.

All students of verbal learning are aware that in the behaviouristic model verbal learning is treated as an associative function. Even in this area a number of investigators have produced phenomena which are not directly explicable in S-R terms even though abstract and unobservable mediational mechanisms are evoked. For, example, the category clustering of Bousfield in which random input comes out as organised output, the chunking of Miller, in which numerous units are recognized into a few larger units, the subjective organization of Tulving (1962, 1966) in which organization of recall becomes unique, to the personal hierarchical organization of Mandler (1967, 1970, 1972) in which some general conceptual structure imposes organization on particular set of items; and finally, the natural language mediation to encoding of presented material as well as string of other organizational processes in memory, illustrate findings which are not easily and convincingly interpreted in associative terms even though the concept of association is stretched to its extreme limits.

Let me take another example to show the extent of tenacity the S-R psychologists have evinced to keep the attention out of the domain of psychological phenomena and concepts. Every student of psychology knows that the controversy between the continuity *vs* non-continuity theories of discrimination learning was centred around the empirical acceptability of the concept of attention. We also know how Spence and Hull and subsequently others laboured to develop and extend the algebraic summation of response strength and stimulus generalization tendency so that cognitive concepts like attention and hypothesis could be kept out of the ambit of experimental psychology. In his well known experiments Lawrence interpreted the obvious attention in terms of a two stage discrimination learning and stressed the concept of the acquired distinctiveness of cues. This idea was further subjected to experimental tests in reversal non-reversal shifts (Kendler & Kendler, 1962). The obtained results required further modification of the theory. The process of fresh experimental findings and further modification of the theory still continues and the S-R theory has become too complicated and circumspect to maintain its elegance and innumerable examples of these types can be added. However, the point I want to drive across is that the learning paradigm proved highly fertile for experimental studies and at the same time required frequent modifications. This process of modification becomes so widespread and frequent that the experimental psychologists become disenchanted with micro or miniature theories. The trend and popularity of micro theories can be easily seen in the area of verbal learning, extinction, STM, LTM, concept learning, aversive learning etc. During seventies, as a matter of fact, the S-R paradigm started crumbling to pieces bit by bit and the movement of cognitivism initiated in the sixties became a powerful movement, unfurling new ideas, areas and experimental procedures. In order to appreciate the aims and emphases, methods and conceptual frameworks of the contemporary experimental psychology it is necessary to look at a new movement initiated by a number of psychologists in the fifties and forcefully launched in 1960 and afterwards.

Three well-known psychologists, George Miller, Galanter and Pribram stayed together at the Centre for Advanced Study in Behavioural Sciences at Stanford, California for a year. They deliberated about the state of affairs prevailing in the then psychology, and come to the conclusion that the then S-R psychology was not acceptable and the S-S psychology of the cognitivists left a big gap between cognitive map and action. They developed a new plan of metatheoretic assumptions and theoretical framework for understanding complex human behaviour. As an outcome of their deliberation, a book was hailed as the starting point of a new constructive era in experimental psychology. They accepted the cybernetic model.

In memory the stimulus encoding theories, information processing at various levels such as sensory, short-term semantic level are being widely employed for generating experimental hypotheses, procedures and data. Set structure and feature comparison models of semantic memory, text processing and remembering etc., are new areas of research. Perception of speech, pattern recognition and reading are taken as challenging experimental problems. Attention including self and environment focus has been restored to its early glorious position.

In eighties the most significant development that has occurred is the paradigmatic shift. The experimental psychologist has moved from the passive reactive and associative living organism to the concept of an active, organising, self-regulating system. Even those learning psychologists such as Bandura, Mischel and Kanfer who started in the learning paradigm, have now moved in this direction and approximate this position to their theorizing. Some of the important tendencies apparent in the literature are as follows:

1. Cognitive variables are considered to have major impact on self-regulation (Dulaney, 1968). Mischel (1973) emphasised the fact that the cognitive representations of stimuli are more impactful than stimuli themselves. Bandura (1978) also makes the same point when he avers that people do not react to stimuli mechanically, they actively process and transform them.
2. The goal directed activities are now conceptually analysed as having altogether different components and variables such as encoding and categorising of events, expectancies about the outcome of their actions, subjective desirability of the outcomes and person's self-regulatory systems and plans (Mischel, 1973). It is easily discernible that the vocabulary is entirely different from the one we are used to. Processing system receiving input from the external environment and creating output in the forms of various types of overt behaviours, rejecting the molecular or molar units of behaviour for analytic processes they advocated the TOTE, (test operate-test-exist) hypothesis and developed the concepts of plan, strategies and tactics, execution and image. However, the boldest step they took was to include human will

and intentions, values and instincts and plans for remembering in their new programmatic psychology.

Cognition, perception, attention, memory, and thinking were subjected to fresh studies for verifying hypothesis derived from information processing models. Most of the experimental psychologists lost their suspicion of the mentalistic concepts. In the contemporary experimental psychology comparative cognition has become a highly fertile and respected topic. It has now been shown that animals such as rats and chimpanzees possess recognition, spatial and serial memory. In this connection the writings of Jerison (1978), Downs and Stea (1973) and Olton (1979) can be seen.

Information processing model has now been applied to almost all areas of experimental psychology. Concept learning as a function of the type of rules, amount of information, irrelevance and redundancy of information, feature dominance etc., has become a widely investigated area. In verbal learning the organisational processes, role of structural redundancy, imaginary grammatical rules are topics of current experimental research.

On the other hand Bandura (1978) posits a self-system in every individual which is a set of cognitive structures providing reference machines and sub-sets of self-functions such as perceiving, evaluating and self-regulating behaviours according to self-prescribed contingencies. He also posits self-observation and evaluation, personal standards and comparison.

Kanfer (1971, 1977), Kanfer and Hegerman (1981) discuss self-regulation in terms of self-imposed behavioural standards, self-evaluation and concepts present a fairly representative flavour of behavioural analysis that is vogue today.

3. In the matter of motivation and reinforcement psychology moved from deprivation and external reinforcement to intrinsic motive-intrinsic rewards

and now to self-reinforcement which is conceptualised as an internal mechanism which is the source of reward when a desired goal is achieved. Such an interpretation provides fairly attractive account of intrinsic motivation. Some psychologists have moved forward and are now positing a control theory which is more closely akin to the cybernatic model. These ideas are not confined to experimental, social and personality psychology but are being fruitfully extended to broader areas of human interest. The paradigm shift has opened new areas of studies in experimental psychology.

References

Anderson, J.P., & Bower, G.H. (1973). *Human Associative Memory*. New York: Winston and Sons.

Bandura, A. (1965). Influence of model's reinforcement contingencies on the acquisition of initiative responses. *Journal of Personality and Social Psychology, 1*, 189-595.

Bandura, A. (1971). Vicarious and self-reinforcement process. In Glaser (Ed.) *The Nature of Reinforcement*. New York: Academic.

Bandura, A. (1978). The self-system in reciprocal determinism. *American Psychologist, 33*, 344-358.

Bower, G.H., & Miller, N.E. (1958). Rewarding and punishing effects from stimulating the same place in the rats brain. *Journal of Comparative and Physiological Psychology, 51*, 669-674.

Deci, E.L. (1975). *Intrinsic Motivation*. New York: Plenum.

Downs, R.M., & Stea, D. (1973). *Image and environment*. Chicago: Aldine.

Dulany, D.E. (1968). Awareness, rules and proportional control: A confrontation with S-R behaviour theory'. In T.R. Dixon, & D.L. Horton, (Eds). *Verbal Behaviour & General Behaviour Theory'*. Englewood Cliffs, N.Y.: Prentice Hall.

Garcia, J., Erwin, F.R., & Koelling, R.A. (1966). Learning with prolonged delay of reinforcement. *Psychonomic Science, 5*, 121-122.

Garcia J., Kimmeldorf, D.J., & Koelling, R.A. (1955). Conditioned aversion to saccharin resulting from exposure to gamma radiation. *Science, 122*, 157-158.

Garcia, J., & Koelling, R.A. (1966). Relation of cue to consequence in avoidance learning. *Psychonomic Science, 4*, 123-124.

Garcia, J., Mc Gowan, B.K., & Green, K. (1972). *Biological Constraints on Conditioning*. New York: Appleton Century-Crofts.

Jerison, H. (1979). *Evolution of the Brain and Intelligence.* New York: Academic Press.

Kanfer, F.H. (1971). The maintenance of behaviour by self-generated stimuli and reinforcement. In A., Jacobs, & L.B. Sachs, (Eds). *The Psychology of Private Events.* New York: Academic Press.

Kanfer, F.H. (1977). The many faces of self control or behaviour modification, changes, its focus'. In R.B. Stuarts, (Ed.). *Behavioural Management: Strategies, Techniques and Outcomes.* New York: Brunner/Nezel.

Kanfer, F.H., & Hagerman (1981). The role of self-regulation. In L.P. Rewar (Ed.). *Behaviour therapy for depression; present status and future directions.* New York: Academic Press.

Kendler, H.H., & Kendler, T.S. (1962), Vertical and horizontal processes in problem solving'. *Psychological Review, 69,* 1-16.

Lepper, M.R., & Greene, D. (1978). *The Hidden Cost of Reward: New Perspectives on the Psychology of Human Motivation'.* N.J.: Elibaum Hills.

Mandler, G. (1967). Organization and memory. In K.W. Spence and J.T. Spence (Ed.), *The Psychology of Learning and Motivation,* Vol. I, New York: Academic.

Mandler, G. (1970). Words lists and categories. In J.L. Cowan (Ed.), *Studies in Thought and Language.* Tuscon: University of Arizona Press.

Mandler, G. (1972). Organization and recognition. In E. Tulving and W. Donaldson (Eds.), *Organization of Memory.* New York: Academic Press.

Martin, E. (1971). Verbal learning theory and independent retrieval phenomena. *Psychological Review, 78,* 314-332.

McGeoch, J.A. (1931). Forgetting and the law of disuse. *Psychological Review, 38,* 352-370.

MacCorqodale, K. & Meehl, P.E. (1954). E.C. Tolman. In W.K. Estes, S. Koch, K. MacCorquodale, P.E. Meehl, C.G. Schoenfeld, & W.S.Verplanck (Eds). *Modern learning theory* (pp. 177-266). New York: Appleton-Century Crofts.

Melton, A.W., & Irwin, J.M. (1940). The influence of the degree of interpolated learning on retroactive inhibition and the overt transfer of specific responses. *American Journal of Psychology, 53,* 173-203.

Menzel, E.W. (1973). Chimpanzee spatial memory organization. *Science, 182,* 943-945.

Miller, George, A., Galanter, E., & Pribram, K.H. (1960). *Plans and Structures of Behaviour.* New York: Holt.

Mischel, W. (1973). Toward a cognitive social learning reconceptualization of personality. *Psychological Review, 80,* 252-283.

Newton, J.M., & Wickens, D.D. (1956). Retroactive inhibition as a function of the temporal position of interpolated learning. *Journal of Experimental Psychology.*

O'Keefe, J., & Nadal, I. (1978). *The hippocampus as a cognitive map*. New York : Oxford University Press.

Olds, J. (1958). Satiation effects in self-stimulation of the brain. *Journal of Comparative and Physiological Psychology, 51*, 675-678.

Olds, J., & Milner, P. (1954). Positive reinforcement produced by electrical stimulation of septal area and other regions of the rat brain. *Journal of Comparative and Physiological Psychology, 47*, 419-427.

Olton, D.S. (1979). Mazes, maps and memory. *American Psychologist, 34*, 583-596.

Postman, L., Stark, K., & Fraser, J. (1969). Temporal changes in interference. *Journal of Verbal Learning and Behaviour, 7*, 672-694.

Premack, D. (1959). Toward empirical behaviour laws: I. positive reinforcement. *Psychological Review, 66*, 219-233.

Premack, D. (1971). Catching up with common sense or two sides of a generalisation: Reinforcement and punishment. In R. Glasser (Ed.), *The Nature of Reinforcement*, New York: Academic Press.

Premack, D., & Prytulak, L.S. (1961). Natural language mediation. *Cognitive Psychology, 2*, 1-56.

Rosenfield, D., Folger, R., & Adelman, H.F. (1980). When rewards reflect competence: A qualification of the over-justice effect. *Journal of Personality and Social Psychology, 39*, 368-376.

Rosin, P., & Kalat, J.W. (1978). Specific hunger and poison avoidance as adaptive specialization of learning. *Psychological Review, 78*, 459-485.

Seligman, M.E.P. (1970). On the generalities of the laws of learning. *Psychological Review, 77*, 406-418.

Shettleworth, S. (1972). Stimulus relevance in the control of drinking & conditioned fear response in domestic chicks. *Journal of Comparative and Physiological Psychology, 8*, 175-198.

Shiffrin, R.M. (1970). Forgetting: trace erosion or retrieval failure? *Science, 168*, 1601-1603.

Tulving, E. (1962). Subjective organization in free-recall of unrelated wors. *Psychological Review, 69*, 344-354.

Tulving, E. (1966). Subjective organisation and effects of repetition in multitrial free recall verbal learning. *Journal of Verbal Learning and Verbal Behaviour, 5*, 193-197.

Timberlake, W., & Allison, J. (1974). Responses deprivation: An empirical approach to performance. *Psychological Review, 84*, 146-164.

Tolman, E.C. (1932). *Purposive Behaviours in Animals and Men*. New York: Appleton-Centry Crofts.

Tolman, E.C. (1948). Cognitive map in rats and men. *Psychological Review, 55*, 199-208.

Tolman, E.C. (1959). Principles of purposive behaviour. In S. Koch (Ed.), *Psychology: Study of a Science* (Vol. 2), New York: McGraw Hill.

3

Culture as a Psychological Construct

Culture is a ubiquitous term and is one of the most frequently as well as indiscriminately used terms in the everyday parlance of educated people, literary writings and social sciences especially in sociology, anthropology and psychology. Whenever one has to make reference to caste or class, religion or ethnicity, geographical location or nationality, this term is glibly employed. It is employed as a hypothetical construct not only for descriptive purposes but also as an explanatory concept. There is no gainsaying the fact that the term "culture" is derived from the Latin word "cultura" which is a verb and means to cultivate or to till. In the English language the word 'culture' is commonly used as a noun, which etymologically means cultivation or tillage—a kind of human activity or behaviour. However, in social sciences, the conceptualisation of culture has been widened, and there are numerous ways in which it has been defined. Let us examine the conceptual status of this term in the social sciences, especially in psychology.

Conceptualisation of Culture

Most of the contemporary anthropologists and psychologists consider culture not as a process but as a product of human activities. There it is regarded as some kind of medium in which human beings grow, develop and become conditioned to operate. According to Segall *et al.* (1998) culture is out there to be studied. Herskovits (1948) defined culture as the man-made part of the environment. According to Cole and Siribner (1974) culture consists of language, dress patterns, housing ways, common beliefs about the world and child-rearing methods prevalent in a society. In 1871, Taylor considered culture as a complex, which includes knowledge, beliefs, art, morals, law, customs and other capabilities and habits acquired by man as a member of society. One can cite many more such definitions. However, in all such definitions culture is conceptualised as some kind of unspecified medium for human development or as a set of contextual variables that have intervening and moderating effects on the encoding of sensory inputs that initiate different mental processes and culminate in evoking various responses and behaviours. As a part of environment, culture refers to historical relics and monuments, art and architecture, music and literature, rites and rituals as well as myths and traditions.

If one goes through the relevant literature in cultural anthropology and cross-cultural as well as cultural psychologies it become0s self-evident that culture is reflected in special components of man-made environment. Garnst and Norbeck (1976) are of the view that culture is a superorganic phenomenon represented by symbols such as words, gestures and postures, and meanings bestowed upon various objects by the human mind. Culture is mirrored in social order, personality and geographical environment. Theodore Schwartz (1978) conceptualises culture in terms of the idioverse or the individualised texture of a culture. According to him, it is no more than that which is shared by all members of society in the same way as personality is confined to that which is unique to the individual. Thus, it is obvious that in this perspective culture is shared way of life. Schwartz views culture as embodied in a set of constructs or representations of behavioural and other classes of events. The confusion regarding the exact nature and features of culture becomes confounded when one reads a definition of culture referred by Hermans and Kempen (1998). Here culture is defined as "a package of ideas, values and practices; as a repertoire of schemas; as a system of meanings, symbols and actions; as a syndrome of beliefs, norms, attitudes and roles; and as a pattern of self-definitions centred around a theme." Is it possible to comprehend the nature of culture according to this long-winded definition?

Because of diverse and vague definitions of culture and a complete lack of agreement as to what is exactly denoted by this term, some scholars have de-emphasised the quest for a universally acceptable definition. In the last five decades, a group of psychologists calling themselves cross-cultural psychologists has emerged. They assume and assert that culture is an all-encompassing context which is out there to be studied (Pandey *et al.*, 1996; Segall *et al.*, 1998). In cross-cultural psychology it is also assumed that culture determines the development of psychological functions. The cross-cultural psychologists consider it as a methodological approach to understand psychological processes. Precisely because of this, culture is treated as an independent variable. It distinguishes divisions of mankind in different categories.

The concept of culture has been further complicated and made more confusing by some psychologists who call themselves cultural psychologists, such as Bruner, Cole, Shweder and a host of others. J.G. Miller (1997) asserts that culture and psychological processes are mutually constitutive phenomena. Shweder and Sullivan (1993) state that culture and psyche make up each other, and create inter-subjective reality through which worlds are known, created and experienced. Thus culture and self become interdependent. Misra and Gergen (1993) consider culture as a resource.

Cultural psychologists consider culture as cumulative product of human thoughts and actions. It provides some kind of medium or context in which members of society grow and develop. Thus the term culture becomes an arbitrarily coined label for laying stress on behavioural and psychological differences among communities, ethnic groups, nations, religious groups and human habitats.

Culture: A Determinant of Psychological Functions

When employing culture as an independent variable, cross-cultural psychologists identify two or more social, national, ethnic or geographical groups and draw samples from them assuming that: (i) the samples drawn have different cultural antecedents; and (ii) these antecedents function as influencing agents, and lead to differential psychological characteristics. Thus, drawn samples are subjected to measurement of some prespecified behavioural characteristics or psychological functions. Employing this research strategy, crosscultural studies have been conducted with the explicit purpose of assessing and specifying the cultural influence on a variety of (i) behaviours, such as motor behaviour, sexual expression and emergency reactions, (ii) basic psychological processes, such as perception, cognition, thinking and conceptual categorisation, etc., (iii) human abilities, such as intelligence, and (iv) a variety of other psychological functions (George, DeVes, & Hippler, 1975). Even today cross-cultural studies are being conducted and research findings are published. However, a critical review of these studies reveals a number of interesting facts. First, significant differences observed between or among the groups are quantitative and there are no grounds to assert that the samples drawn from the so-called different cultures are representative or can be considered to have even illusory semblance of the required representativeness of the cultural universe. Thus, without authentic and valid measures of the intra-cultural variance, the observed cross-cultural differences are of no consequence. Second, there is not a single study showing some psychological phenomenon present in one culture not being in others. It is but natural to assume that if there are different cultures providing medium of psychological growth, some cultures would give rise to the emergence of some culture-specific phenomenon. However, the empirical findings belie such an assumption.

Keeping in view the problems in cultural and cross-cultural psychological studies, Spiro (1978) has pointed out that cultural determinism is based on three fallacies. First, the fallacy of assuming that cross-cultural variability in social behaviour and personality implies that the human organism is an empty black box because despite observed variability in culture, human beings have a variety of invariant psychological characteristics. Second, the fallacy of confusing a culture with culture in the expression of cultural determinism. Third, the cultural and cross-cultural psychologists are victims of the fallacy of not distinguishing between the phenotypical (observed variability in psychological characteristics) from the geneotypical (underlying genetic make-up). In other words, they have failed to distinguish between the surface structures and the deep structures.

Newly Emerged Challenges

The validity of cross-cultural and cultural psychology has been further challenged by some newly emerged trends on the world scenario. Hermans and Kempen (1998) have brought out some irrepressible challenges to the theory of cultural dichotomies as determinants of behaviour and human personality. These authors point out that in an increasingly interconnected global society, the conceptualisation of independent, coherent and stable cultures becomes totally irrelevant. They have cogently argued that cultural differences in the form of dichotomous distinctions do not meet the challenge of globalisation and its implications for a psychology of culture and self. First, the pervasive influence of cultural contacts and connections owing to globalisation in all aspects of educational, social, political and economic life on this terrestrial sphere, the emergence of cultural mixtures and hybridisation have become an order of the day. Such a trend has undermined the idea of cultures as internally homogeneous and externally distinctive phenomena. Second, it is a crucial fact that even apparently or phenotypically opposite cultures have become integral components of one world civilisation. In modern times, cultures conceptualised as mutually distinguishable behavioural contexts are no more tenable. Finally, it is important to note that the increasing complexity of cultural aspects of life has rendered the idea of cultural distinction and dichotomisation quite problematic.

Blind Spots in Cross-cultural and Cultural Psychology

During the later half of the last century easy availability of fast and efficient travel facilities all over the world and large-scale programmes for exchange of scholars a large number of psychologists visited foreign countries and stayed there for months and years. Naturally enough, they found the ways of life, social interactions, art and architecture, music and folk songs, historical monuments and relics of the host countries very different and strange in comparison to their own. They were not only wonder-struck but got deeply interested in studying these diversities. This was the background in which cross-cultural psychology was formally started and subsequently cultural psychology came into existence.

Thinking and theorising efforts of such psychologists appear to be deeply rooted in the paradigm of environmentalism. As a consequence, cross-cultural psychologists conducted their empirical research with three metatheoretic assumptions, that is, (i) culture is out there to be studied, and is a medium in which psychological characteristics grow, (ii) culture as a medium or context functions as a set of antecedent conditions giving rise to psychological functions and phenomena as consequent conditions, and (iii) empirical differences obtained from cross-cultural comparisons are the critical facts for theorisation.

The aforesaid assumptions are grossly biased and untenable. To assume that culture and psychology are mutually constitutive phenomena, and that there are fundamental differences in human thoughts and cognitive processes across cultures, is in contradistinction with some well-established findings of cultural anthropologists. It is now undeniably true that there are cognitive universals. The reasoning and thought processes in different cultures do not differ, and the structural features of language, such as, words, organised sequence and rules are shared by all languages (Cole & Scribner, 1974). On the other hand, the assumption stemming from the once popular behaviouristic environmentalism that culture as context determines behaviour and psychological functions now stands discredited. Moreover, significant findings by socio-biologists about genetic influences on mental abilities, personality development and a variety of social behaviours have further strengthened the principles of biological evolution in all branches of the life sciences. Now Herbart Spencer's (1850-1900) dictum is universally accepted. It states that a single cosmic principle of evolution predisposes human beings in their mental functions; and behaviours change in a definite direction, that is, from simplicity to complexity and from homogeneity to heterogeneity. In this backdrop to relegate this biological principle to the background and to reject the unity in human nature is not only unscientific but also provides unintentional support to those forces that are interested in keeping mankind divided in different parts of the world.

Now let us look at the conceptualisation of culture as a man-made context or medium of mental growth and functioning. Such a conceptualisation makes culture a man-made product. In doing so one commits the error of ignoring the distinction between products and processes, or between manifest forms and latent substance. Every product or form is a consequence of some specific process or processes. All products and forms including cultural ones, such as, myths and mores, rites and rituals, and art and architectures undergo modification, but the processes underlying them remain unchanged. Another point worth mentioning is that the same processes may give rise to different products and forms depending on extraneous factors. For example, the psychological processes underlying the method or style of building houses of worship are the same but buildings appear to be different. The apparent differences are of little importance in understanding the psychological processes unless the ingroup-outgroup syndrome becomes operative in such comparisons.

In English dictionaries (Webster's and Oxford) the word culture is both a noun as well as a verb. Thus, culture implies both, that is, some kind of action or activity, as well as an objective entity or product. However, cross-cultural and cultural psychologists have been totally anchored on culture as a noun and have disregarded the verb or process aspect of culture. There is no gainsaying the fact that cultural forms vary from society to society, country to country and also across different time periods, but the processes underlying them are unchangeable and universal.

Eversince psychology was recognised as a science in countries other than those in which it originated and developed, tests, tools and study materials were imported, translated and adapted to suit the local conditions and languages, and were used for further research. Segall *et al.* (1998) considered this as one of the three complementary goals of cross-cultural psychology. New tests, tools and experimental materials were also developed in accordance with the requirements of scientific methodology. It is quite correct to call such tests indigenous. However, Kim and Berry (1993) unwittingly initiated a very harmful trend by coining the term "indigenous psychologies". It is but natural to assume that indigenous psychologies are those that are country or culture-bound, and employ not only indigenous tests, tools and material but also require indigenous concepts and constructs, methods, theories and data. Thus this term has encouraged many psychologists in universities, especially located in Asian countries, to imagine the existence and reality of culture-bound psychological phenomena and principles. Is it not amusing to find that the world has not needed indigenous Botany or Zoology or Genetics despite the fact that the flora and fauna of different territories are different in form and functions? Talk of indigenous psychologies will, or is likely to, retard the growth of scientific psychology.

Culture: Group's or Individual's Property?

Dichotomisation or categorisation of culture entails the tacit assumption that culture is a product that provides context or medium for personality and behavioural development. It is under this assumption that people talk of Western and Oriental cultures, American, English, French, Arab, German, Chinese, Japanese and Indian cultures. Quite often cultures are differentiated from one another on religious grounds, such as Christian, Muslim, Hindu, Jewish and Shinto cultures. On the basis of habitats also cultures are categorised as rural and urban. Some of the ways in which culture has been categorised in cross-cultural studies are based on nationality, geographical locations and developmental level—tribal and non-tribal, native born and immigrant citizens of a country. It is worth noting that criterial ambiguity in categorisation leads to innumerable ways of categorising cultures and creation of taxonomic chaos.

This point can be illustrated by taking the case of Indian culture. If one endeavours to differentiate the distinguishing features of Indian culture, one is bound to get lost in the attractive alleys of the phrase "unity in diversity" or "Ganga-Jamuna Sanskriti". The supra ordinate Indian culture consists of as many basic cultures as there are officially recognised languages or dialects. That was the reason that in mid-1950s States were reconstituted. The States of Uttar Pradesh, Bihar, Jharkhand, Bengal, Assam, Maharashtra, etc., are believed to have different cultures. Then there are tribal and non-tribal cultures, rural and non-rural cultures. Caste and class, religions, vocations, joint and nuclear families, etc., are also employed for categorisation of cultures. As a matter of fact, if one delves deep into the lifestyles, belief systems, social interactional patterns, one may easily find two neighbouring families belonging to the same class and caste having diametrically opposed cultural patterns and practices, cognitive styles and personality characteristics. Likewise it is also easy to find four brothers having identical genetic roots, upbringing and socialisation ethos but different belief systems, ethical values, cognitive styles and biases. On the basis of such incisive observations, it is reasonable to conclude that neither the conceptualisation of culture as context of psychological development nor the theory of cultural determinism is a valid way of studying culture.

In this backdrop of confusion created by the metatheoretic assumptions of cross-cultural psychologists, the most promising wayout calls for focusing attention on culture as a process or activity indulged in by human beings. These processes just like perceptual, thought or memory processes have to be considered as individuals' property. It is only then that one can really understand some individuals being more cultured, cultivated and refined than other individuals. Acceptance of this perspective for conceptualising culture will lead to realisation of the dream of Munroe and Munroe (1997) that "cross-cultural psychology will be shown to have succeeded when it disappears".

Culture as Psychic Potential Manifested in Human Behaviour

Culture, cultural processes and products/forms will be best understood and crystallised if they are contrasted with "nature", natural processes and products. Nature consists of physical objects such as earth, mountains, deserts, rivers, oceans, and stars and planets, biological entities, such as, flora and fauna, and a variety of elements. These objects, entities and elements are active, reactive and interactive according to the laws and principles of nature and give rise to variant climates, infinite kind of flora and fauna ranging from one celled amoeba to the structurally far more complex Homo sapiens. Human beings also like any other living animal are born, grow and survive in accordance with the natural laws and principles. However, human beings are also unique because they are born with genetically determined, in-built and innate potentials to grow in such a way that they acquire capacities to transcend the constraints of natural laws and processes. One of these potentials is concerned with cultural processes and provides latent capacity for acting upon natural objects, modifying or transforming and refining them with a view to making survival easier, more comfortable and pleasurable hedonically as well as mentally.

It must be kept in mind that in human beings natural and cultural processes as well functions are like antonyms. Natural processes are spontaneous exhibiting the principle of least resistance and are impulse guided. On the other hand, cultural processes are deliberate, requiring best effort and guided by imagination and constructive thoughts. Wherever there is interaction between man and nature, man endeavours to quell the natural processes and subjugate them to cultural processes. Eating, mating, seeking safety and living amidst fellow human beings are natural activities for survival, but cooking spiced and well-flavoured, tasty food, searching out clean drinking water, getting married to a member of the opposite sex and living together with mutual trust and fidelity, making jams and marmalades of fruits, building climate suited and weather friendly, comfort providing and visually pleasing houses, converting the acoustic properties of nature into musical systems, etc., are examples of cultural activities and processes. In social interaction, to be selfish and guided by one's needs and feelings is illustrative of natural processes, while being sensitive to the needs and feelings of others, showing tolerance and compassion, indulging in and encouraging prosocial and proactive behaviour and having positive thinking, etc., exemplify cultural processes.

Cultural activities and behaviours of individual members in all communities and countries are characterised by the same kind of individual differences as observed in the assessment of other psychological processes. These individual differences are jointly determined by (i) genetic make-up or heredity which determine the strength of cultural potentials, (ii) psychic factors, such as one's upbringing, familial interaction, socialisation, schooling and unique developmental experiences, which provide opportunities for acceleration or retardation of the manifestations of cultural potentials, and (iii) environmental factors which provide resources and direction for cultural operations leading to varied forms of cultural products. Thus, in spite of observed differences in cultural products of all communities and countries of the world, the processes and psychological forces that underlie them are the same. They can be better understood if treated as dependent variables like any other basic psychological process.

References

Cole, Michael & Scribner, Sylvia (1974). *Culture*. New York: John Wiley.

Garnst, Frederick, & Norbeck, Edward (1976). *Ideas of Culture: Sources and Uses*. New York: Holt, Rhinehart.

George, A., De Vos., & Hippler, Arthur E. (1975). Cultural psychology: Comparative study of behavior. In G. Lindzey, & E. Aronson, C. (Eds.), *Handbook of Social Psychology* (Vol. 4) (pp. 323-417). New Delhi: Amerind.

Hermans, Herbert J.M., & Kempen, Harry J.G. (1998). Moving cultures: The perilous problems of cultural dichotomies in a globalizing society. *American Psychologist, 55*, 1111-1120.

Herskovtts, M.J. (1948). *Man and his Work: The Science of Cultural Anthropology*. New York: Knopf.

Kim, U., & Berry, J.W (Eds.) (1993). *Indigenous Psychologies: Research and Experience in Cultural Context*. Newbury Park, C.A.: Sage.

Miller, J.G. (1997). Theoretical issues in cross-cultural psychology. In J.W. Berry, Y.H. Poorting & J. Pandey (Eds.), *Handbook of Cross-cultural Psychology: Theory and Methods*. Boston: Allyn & Beacon.

Misra, G., & Gergen, K.J. (1993). On the place of culture in psychological sciences. *International Journal of Psychology, 28*, 225-243.

Munroe, R.L., & Munroe, R.H. (1997). A comparative anthropological perspective. In J.W. Berry, Y.H. Poorting, & J. Pandey (Eds.), *Handbook of Cross-cultural Psychology*. Boston: Allyn & Becon.

Pandey, J., Sinha, D., & Bhawuk, D.P.S. (Eds.) (1996). *Asian Contributions to Cross-cultural Psychology*. New Delhi: Sage.

Schwartz, Theodore (1978). *Where is the Culture: Personality as the Distributive Locus of Culture in the Making of Cultural Anthropology,* Berkeley and Los Angeles: University of California Press.

Segall, Marshal, Lonner, Walter J., & Berry, J.W. (1998). Cross-cultural psychology: A flowering discipline. *American Psychologist, 55,* 1101-1110.

Spiro, Melford E. (1978). *Culture and Human Nature in the Making of Cultural Anthropology,* Berkeley and Los Angeles: University of California Press.

Sweder, R.A., & Sullivan, M.A. (1993). Cultural psychology: Who needs it? *Annual Review of Psychology, 44,* 497-523.

Taylor, E.B. (1871). Cited in Chapter 1. In Michael Cole, and S. Scribner (1974). *Culture,* New York: John Wiley.

4

Some Reflections on Contemporary Psychology

Ever since psychology acquired independence as an empirical science, it has continually grown and developed under a variety of metatheoretical assumptions, paradigms, and problematics. It gained academic recognition not as an unitary branch of knowledge but as a multi-streamed intellectual endeavour divided into a variety of schools, such as structuralism, functionalism, gestalt approach, hormic psychology, psychoanalysis, and behaviourism. However, with the emergence of neobehaviourism, and the popular appeal of learning paradigm, some of the schools either became extinct or were amalgamated, while others were marginalized in the main-stream of behaviouristic experimental-theoretical psychology. At this stage, the ontological question concerning the mind-body relationship was successfully evaded by banishing the mind and mental activity from the realm of psychological studies as something mysterious, unscientific, and irrelevant. It was boldly assumed that 'man' was a passive-reactive entity and its behaviours were completely environment-determined. Such an approach, largely guided by the positivistic view of science, attained the pinnacle of success and glory under the supremacy of Hullian hypotheticodeductivism, on one hand, and to Skinner's positivistic inductivism, on the other. A number of psycho-analytic concepts and constructs were appropriated by the neo-behaviourists, and the cognitivistic contributions of Tolman were redefined and incorporated in the behaviouristic mode of theorization.

After the Second World War, a variety of phenomenal developments in the physical and biological sciences, especially cybernetics and information theory, along with a variety of critical experimental findings in psychology laboratories inspired a paradigm shift. The exiled mind and mental processes were brought back into psychological theorization under the aegis of cognitivistic approach. The ontological question of mind-body relationship was resolved by accepting materialistic monism in the form of a man-machine metaphor. The unobservable but objectively explorable events occurring between the input and output are now subjected to experimental investigations and the cybernetic organization or structure is accepted as the core idea in the cognitivistic approach.

The contemporary psychology with its cognitivistic mainstream and other marginal approaches such as humanistic, neo-psycho-analytical and socio-biological interpretations of human behaviour has been, and still is, subjected to scathing criticism. In the beginning, issues raised were endogenous and that too within the accepted metatheoretic framework. Complete absence of long range theoretical systems, mushroom growth of mini-theories for explaining highly specific behavioural phenomena, emphasis on methodological sophistication at the cost of intellectual substantiveness, lack of cumulativeness in research findings, insensitiveness of research findings *vis-à-vis* real life psychological problems were some of the charges put forward against the academic psychology. These charges were and still are basically constructive and call for improvement in the existing theorization and for concentration on relevant research, but do not threaten the survival of psychology as an empirical science. However, during the last few years a variety of exogenous issues have been raised, which are likely to destroy the long standing metatheoretic assumptions and generally accepted problematics of the contemporary psychology. These issues require serious discussion.

Exogenous Criticism

As a natural corollary of intellectual development in the post-positivistic philosophy of science, two planks of attacks against science in general and psychology in particular have emerged.

Basically they are the issues of scientific objectivity and value-neutrality. It has been forcefully argued that scientists, and much less the social scientists can maintain value-neutrality of sciences is untenable. Scientists are guided by values in selection of research problems, especially in applied area, and the kind of concepts, constructs and theoretical formulations they put forward in the process of disseminating knowledge affect existing social values and promote newer ones. Gergen (1991) has correctly pointed out that there is "no means of separating 'is' from 'ought'". Psychologists, while engaged in psychopathological research and practising psychotherapies, or those studying social problems and endeavouring to promote quality of life, or those investigating behavioural and mental problems of children with a view to facilitating and promoting the objectives of educational programmes, are indeed guided by prevalent social values. Such value-oriented researches and practices of psychologists or social scientists do not impair the scientific nature of their work. Thus in the contemporary context of scientific development, value-neutrality does not exist and has become a non-issue.

The second plank of attack focuses on the empiricist foundationalism of modern psychology. Gergen (1991) points out that there is no theory of rational induction. In the inductive mode of theorization, one moves from observations or empirical facts or data to theoretical conceptualization. Here the basic question is: How do inductivists select what is to be observed and what is not to be observed? Even if one asks purely exploratory questions about human behaviours, hidden behind such questions are inarticulated hunches or expectations. Such hunches and not the observations provide observational facts for empirical generalizations and theoretical formulations. Hanson (1958) has reasoned that raw observations are incapable of providing theoretical positions. Long back in 1938, Reichenbach made a distinction between the context of discovery and the context of justification and this distinction has been forcefully used by Popper (1968). He reasoned that the context of justification was subject to rational rule of procedure, but the context of discovery has to remain a mystry in the positivistic inductivism. Acceptance of such a position has naturally hastened the conclusion that theoretical explanations cannot be derived from sense-data. Scientific description made on the basis of observation is thoroughly saturated by theoretical standpoint. Thus, there is no rational induction, and inductive descriptions are in essence deductive.

In this context it is relevant to look at the deductive mode of theory construction and its empirical verification. It is well-known that in this mode of theory construction, one starts with some hypothetical propositions or postulates and designs empirical investigations to test their empirical validity. These postulates belong to the context of discovery. Now, the crucial question is: How and where from these postulates emerge? It is not enough to say that they emerge from ambiguous processes of creative intuition. Long back, experimental psychologists had unequivocally shown that the manifestation of insight was rooted in past experiences and productive thinking. Observation is experiencing. Thus in substance, it does not matter whether one is engaged in inductive or deductive mode of theorizaiton; past observations formally designed or casually experienced are characterized by regularities and lawfulness, and give rise to hypothetical propositions or inarticulated hunches. Carefully planned further observations are the only dependable means of verification or falsification of such propositions.

At this stage, it is imperative to analyze observations leading to theorization in the light of the criticism launched against empiricism by the post-modernists, post-structuralists, relativists and social constructionists, especially Gergen (1987). It has been stated that theories about the world are not representational or mimetic, but the result of human interchange or construction. Social constructionists do not consider scientific knowledge as mirrors or a map of independent reality. To me it appears that to accept such statements is exactly getting entrapped in a new form of solipsism. Language, social negotiations, rules of procedure, literary tropes and rhetorical stratagems do not shape the perceived or observed reality. Reality, be it physical, social or psychological, has independent existence and human senses are sensitive to them to a great extent. Sense organs provide informational input for mental activities and words reflect these activities.

Theoretical problems arise if and when one tries to analyze the representational aspects of mental activities in the framework of man-machine metaphor. Haugeland (1978) has analyzed the man-machine metaphor and distinguished three features that are characteristics of the machine that minds. First, mind is a system composed of subsystems each having a set of states and state-transitions. Second, it functions in such a way that it is describable in terms of representations of its inputs and outputs. Third, each subsystem has representations and rules. Thus machine that minds has perceptual-cognitive capacities (Kurtzman, 1987). In spite of the prevailing controversy about the representational nature of mind, there is no gainsaying the fact that mind apprehends reality and the empirical method of gaining knowledge of reality is the only self-correcting method available to man. How is knowledge represented in the mind is a question to be resolved by way of fresh theoretical formulations and empirical scrutiny. To dispense with the empirical method is to end the possibility of a science of psychology.

Empirical Method and Endogenous Issues

Most of the endogenous issues raised against the contemporary psychology call for a fresh review of the use of empirical method in behavioural and mental research. The mainstream psychology has not developed the way it should have because of researchers' deep-rooted proclivity to treat developments in physical and biological sciences as their models of research and theory construction. Harre, Clarke, and De Carlo (1985) have called it the tendency of scientism. Psychologists have not only borrowed vocabulary from natural sciences, but have got involved in 'taking over methods or techniques of enquiry that have grown up in the sciences of inanimate nature' (Harre *et al.*, 1985, p. 5). On the other hand, psychologists have also committed the error of treating their subject matter of study, *i.e.*, behaviour and mental process, in the way natural scientists have treated physical, chemical, plant and animal structures and functions. It is precisely because of this reason that the data are too artificially collected to be reflective of real life behaviours and mental activities.

While devising empirical techniques psychologists so far have not seriously taken into account the fact that the subjects or respondents are more or less the same and are as observant as the researchers are, of course, with different hunches and expectations. It is largely because of such non-realizations that research data get confounded by a variety of experimental errors (Rosenthal, 1966). Another pitfall of the current researches is that there is more emphasis on group data than on individual data, notwithstanding the fact that in clinical practice and under the aegis of operationism single subjects are assessed and studied.

With the advent of cross-cultural investigations and emergence of cultural psychology, the concept of universal psychology has been almost abandoned, but there is the danger of too much reactance to universalism, and creation of too many cultural or indigenous psychologies. Human beings are the same across all cultures and countries to the extent they have indentical anatomical and neurophysiological structures. It is reasonable to infer that in basic mental processes such as attending, perceiving, learning, memory and thinking all human beings are similar but in products they differ to the extent they are experientially, ecologically and culturally socialized in different ways. Thus, for studying basic processes theoretical constructs can be universal while for studying other psychological functions, theories and techniques have to be indigenous. Under this assumption indigenous issues can be satisfactorily resolved. Thus what is needed is a shift in paradigm and problematics of psychology.

A Modified Paradigm

If psychology has to continue as an empirical science, investigations and theory construction have to be carried out within a modified set of metatheoretical assumptions and psychological problematics have to be redefined. As a matter of fact, a number of writers have cogently pleaded for enlarging and modifying the existing paradigm. In this context the names of Harre (1987) and Carver and Scheier (1981) are worth drawing our attention. The way Harre has talked of enlarging the paradigm envisages a complete breakaway from the prevalent psychological knowledge because he has discovered 'numerous be-setting sins of scientific psychology' and higlighted the virtues of ordinary language analysis, paying explicit attention to the local moral order. The paradigm suggested by Carver and Scheier proposes a new orientation in the light of the shortcomings of the learning paradigm. However, their commitment to the cybernetic model of man and minding has restricted the scope of their metatheoretical assumptions.

The present author very strongly feels that the prevalent scientific psychology has provided innumerable insights into the nature of human behaviour and the psychological processes underlying them and those insights can be and should be fruitfully utilized in modifying the metatheory and problematics for future psychology. It is in this backdrop that the author has the audacity of suggesting a fresh set of metatheoretical assumptions. The author is of the view that the man-machine metaphor is the best way of conceptualizing mental activities. Of course, the concept of machine itself is a drastically changed entity from the one as is commonly understood by most of the psychologists today.

1. *Human behaviour and mental activities or processes that underlie them are basically developmental in nature, and primarily originate and grow in a strictly deterministic-mechanistic manner, and with the emergence and development of self they tend to become rule guided and self-regulated.*

This assumption is suggested by a variety of research findings in different areas of research such as origin and development of self, cognitive social learning and self-reinforcement (Bandura, 1976; Mischel, 1973). The notion of 'plan' and TOTE hypothesis developed by Miller and others (1960), Lewin's notion of inter-personal-tension system, Freud's enunciation of the emergence or differentiation of ego and superego from the primordial id, Piagetian theory of cognitive development and a few others such as these, reveal enough pointers for justifying this assumption. To be more explicit in this respect, a number of interesting dichotomous processes that have been discovered in different areas of psychological research can be cited to strengthen the foundations of this assumption. It has been noted that selective attention can be directed outward to the environment or directed inward to the self. Carver and Scheier (1981) have very elaborately demonstrated the differential functioning of the two. In causal perception, external and internal attributions are well known. The concepts of field dependence and field independence, internal and external control, extrinsic *versus* intrinsic motivation, private and public selves and many more such dichotomies give important leads for studying developmental processes under this assumption.

2. *The extent to which human beings have identical sensory, neurophysiological and glandular structures, the basic psychological processes especially those elicited and maintained by stimulus determinants are universal, but the development of self, including behavioural standards, belief systems and behavioural strategies is determined by socialization patterns, social settings, ecological conditions and constanty changing rituals and rites.*

The above assumption is also rooted in the achievements and failings of the scientific and cross-cultural psychology as well as in empassioned appeals that are being made for the need of indigenous psychologies. It is a matter of common knowledge that the basic psychological processes such as selective attention, vigilance, perceptual organizations, illusions, three-dimensional perceptions, stimulus encoding, storage and retrieval processes, concept learning, reasoning errors, problem-solving strategies etc. exhibit universally observable features. It is only in the group, social or organizational settings that behavioural standards, belief systems, motivational patterns and goals show great variations in different societies and ecological conditions.

3. *Psychological development through life span from stimulus-regulation to self-regulation is neither autonomous nor spontaneous. The experiential qualities and quantities, socialization patterns and cultural settings act as facilitators or retarders and any aspect of self may become stunted, debilitated or sharpened at any stage of development.*

Such an assumption is forcefully suggested by a variety of studies of perceptual learning, cognitive functioning and problem-solving behaviours in natural habitats such as deprivational or chronically poverty stricken environments, or different types of ecological settings. Developmental studies of cognitive, motivational and personality functions in diverse types of social settings and the emergent controversy between the deficiency *versus* difference hypothesis is an important source of this assumption. If self is conceptualized as multifaceted entity consisting of more or less coherent representations stored in memory its developmental nature has to be necessarily accepted. During various developmental stages, manifestations of behavioural-mental problems such as learning disability, hyperactivity, unsocialized aggressive reactions, learned helplessness, social dependence and innumerable other behavioural features provide fruitful grounds to be tested within the framework of this assumption.

4. *On one hand, because of the uniqueness of the subject matter of study, psychology has necessarily to be an empirical-analytical science both at the microcosmic as well as the macrocosmic levels, and on the other hand, because of the inseparability of human from moral-ethical saturations, it has to be conceptualized, concretized and promoted as an action-facilitative technology of behaviour both at the individual and collective levels.*

Uniqueness of human in general and of each individual in particular is not a new idea. Uniqueness of human personality was conceived by Adler who considered each individual as a unique configuration of motives, interests, values and traits. Gordon W. Allport (1961) observed that 'the outstanding characteristic of man is his individuality. He is a unique creation of the forces of nature' (p. 4). He has elaborated on the uniqueness of personal dispositions. Since experimental-theoretical psychologists were and still are basically interested in universal laws of behaviour and mental activities, they believe that uniqueness could be accounted for in terms of differences in the compounding of the same underlying mental processes. That is how uniqueness of human beings is swept under the carpet of error variance. Even psychology of individual differences is not free from this shortcoming. What is needed is to lay greater emphasis on within subject experimental studies, so that psychological findings may become more meaningful at the microcosmic level. In clinical and counselling practices each client is treated as a unique case, and the psychologists of today have to re-invent uniqueness and develop more effective techniques of discovering and measuring unique behavioural-mental features. In brief, as an empirical-analytical science, psychology has to recognize the importance of microcosmic studies.

Since moral ethical values involved in human behaviour, especially in social settings, are inseparable in scientific studies and their vulnerability to be affected by theoretical formulation, it is essential for psychologists to accept the social and cultural ideals and study the retarders and impediments that hamper the growth of healthy self-regulation. Realization of the hidden self-potential, self-control, self-understanding and self-broadening are universal ideals. It is in this backdrop that psychology has to develop behavioural and mental techniques for removal of retarders in the path of, and for promotion of, self-regulation. At this point, a note of caution is necessary. Psychology in order to be actional-facilitative does not imply in anyway that psychologists have to assume the role of social reformers. Rather, they have to be like medical scientists and practitioners as well as mental health consultants.

References

Allport, G.W. (1961). *Pattern and Growth in Personality.* New York: Holt.

Bandura, A. (1976). *Social Learning Theory.* Englewood Cliff, NJ: Prentice Hall.

Carver, C.S., & Scheier, M. (1981). *Attention and Self-regulation: A Control Theory Approach to Human Behaviour.* New York: Springer-Verlag.

Gergen, K.J. (1987). On the social construction of knowledge. In Esthues (Ed.), *Basic Principles for Social Science in Our Time.* New York: University of St. Jerome's College Press.

Gergen, K.J. (1991). Emerging challenges for theory and psychology. *Theory and Psychology, I,* 13-35.

Hanson, N.R. (1958). *Patterns of Discovery.* Cambridge: Cambridge University Press.

Harre, R. (1987). Enlarging the paradigm. *New Ideas in Psychology, 5,* 3-12.

Harre, R., Clarke, D., & De Carlo, N. (1985). *Motives and Mechanisms: An Introduction to the Psychology of Action.* London: Methuen.

Haugeland (1978). The nature and plausibility of cognitivism. *Behavioral and Brain Sciences, 2,* 215-226.

Kurtzman (1987). Deconstruction and Psychology: An introduction. *New Ideas in Psychology. 5,* 33-71.

Miller, G.A., Galanter, E., & Pribram, K.H. (1960). *Plans and the Structure of Behavior.* New York: Holt.

Mischel, W. (1973). Toward a cognitive social learning reconceptualization of personality. *Psychological Review, 80*, 252-283.

Popper, K.R. (1968). *The Logic of Scientific Discovery.* New York: Harper & Row.

Rosenthal, R. (1966). *Experimenter Effects in Behavioural Research.* New York: Appleton-Century-Crofts.

5

Knowledge and Faith in Contemporary Psychology

Notwithstanding the fact that 'knowledge' and 'faith' are two different words of English language, having two different connotations and denotations, social scientists in general and psychologists in particular often blur the difference, and consider the two interchangeable. According to the Illustrated Oxford Dictionary, the word, 'knowledge', as a noun means (i) familiarity gained by experience; (ii) person's range of information; (iii) theoretical or practical understanding (of); and (iv) the sum of what is known. According to the New Webster's Dictionary, 'knowledge' means (i) acquaintance with facts, truths, or principles as from study or investigation; (ii) familiarity, conversance with a subject, language or branch of learning; (iii) acquaintance with a thing, place, person at sight, experience or report; (iv) the fact or state of knowing; (v) perception of fact or truth; (vi) clear and certain mental apprehension; (vii) the state of being cognizant or aware of a fact or circumstance; (viii) practical understanding of an art or skill; and (ix) the body of truths accumulated by mankind in the course of time. It is worth taking note that the referents of 'knowledge' according to these two dictionaries are *experience, information, acquaintance, conversance, perception, cognizance and understanding*. These referents of knowledge of any entity or event specify as to what it is, and what are its discriminable features.

The word 'faith', according to the Oxford Dictionary, means (i) reliance, trust in... (ii) beliefs in religious doctrines especially such as affect, character and conduct, spiritual apprehension of divine truth apart from proof; (iii) system of religious belief;

(iv) things to be believed; and (v) warrant, promise, engagement, loyalty, fidelity. According to the New Webster's Dictionary, 'faith' means (i) confidence or trust in a person or thing; (ii) loyalty, fidelity to a person, promise or commitment; (iii) belief not substantiated by proof; (iv) spiritual acceptance of truth or reality not certified by reason; (v) belief in God; and (vi) belief in the doctrines or teachings of a religion. In nutshell it may be asserted that faith refers to acceptance of a proposition without questioning its veracity because others, who are more experienced, superior in position or command authority have stated so. Once faith in the veracity of some proposition develops one holds on to it tenaciously.

Origin and Development of Knowledge and Faith

Thus, it is clear that knowledge and faith refer to two states of cognition different from each other. However, they originate from the same source, and fulfil the same kind of human needs. It is now well known that human beings have a variety of needs, motives or drives which arouse everyone to become active for fulfilling those needs. The need for knowledge of the environment and self is one of such needs. Psychologists of diverse theoretical backgrounds and research interests have analyzed human needs in different ways, but majority of them have come to the conclusion that the need for knowledge and understanding is a basic motivational force.

A long time ago Freud (1922), realized that the hedonism or the pleasure-pain principle was not the only motivational force sufficient to account for all kinds of human activities. Human beings indulge in a variety of activities for understanding oneself. He, therefore, posited another motivational force, which he called, the instinct of mastery.

Festinger (1954), while in the process of formulating his theory of social comparison, asserted that each individual is endowed with a drive to evaluate his cognition (opinions, beliefs, sensory impressions etc.) of the environment in which one lives, and also to assess one's capabilities for controlling the environment. He has argued that this drive may be subsumed within the framework of hedonistic principle because information seeking from one's environment and appraisal of one's capabilities may be rewarding or punishing in the long-term. In the present context it is not relevant to argue in favour of, or against, this proposition. It is a well-known fact that humans are information seekers as manifested in the form of curiosity, and also self-evaluators so that they may enhance their environmental understanding and self-appraisal capability.

Maslow (1943-1967), one of the pioneers of humanistic psychology, postulated a set of growth needs called self-esteem, competence, recognition or mastery, and self-actualization. For recognition in society one must have mastery over some facets of environment. Lack of mastery creates a feeling of inferiority and leads to denial of social recognition and appreciation necessary for self-enhancement. According to Maslow it arouses humans to be curious and seek information by observing things and events, and then inferring their causes, or seeking other's opinions, reading books or by using any other means, including the method of authority.

After analyzing naive psychology and cognitive processes involved in it, Fritz Heider (1958) posited two motivational forces, *i.e.* (1) the motive or need to form a coherent understanding of the world, and (2) the need to control the environment. It is because of these needs that every person tries to encode and interpret all sensory stimulations of which one becomes conscious. In this process one endeavours to understand the what, how, and why aspects of sensory inputs. One of the essential conditions for satisfying this need is acquiring ability to predict how environmental objects/events would change and how others would respond to it. The implicit assumption underlying it is that all events in the world are lawful, and that specific antecedent conditions are followed by specific consequent conditions. This assumption leads to causal perception or attribution.

Keeping in mind the aforesaid theoretical framework, the concepts of knowledge and faith are indistinguishable because both are acquired by common human beings with the help of a number of methods, such as, causal perception, seeking opinion of others, edicts of elders and authorities, tenacity, accumulated subjective experiences, and heuristic reasoning etc. These methods for understanding objects and events in one's environment are learned by children in the course of their upbringing in all societies and all over the world. Because of these methods knowledge and faith remain indistinguishable.

Distinctive Features of Knowledge

If one takes hints from the dictionary meanings of the two terms, it is reasonable to assert that knowledge, as a cognitive state refers to understanding of the what, how and why of a set of sensory inputs received from environment. This kind of understanding fulfils the requirements of three criteria, *i.e.* true explanation, control, and conditional prediction. Ever since its beginning as a science, psychology accepted physical and biological sciences as its model, and has now become an integral part of cognitive science. Psychologists explore functional relationship between specified antecedent and consequent conditions through naturalistic as well as controlled observations. Such empirical methods have been chosen because they are *self-corrective*.

It is the self-corrective character of the empirical method that has led to drastic overhauling of the behaviouristic paradigm. Robust empirical proofs accumulated over the years have proven the existence of a variety of mediational or cognitive processes that intervene between the information input and behaviour as output, through active nature of mind. It is now indubitably proven that perception is not a photocopy of the informational input but an informed construction (Kelly, 1955). Psychologists themselves using empirical method have brought to light numerous confounding factors that vitiate the contexts in naturalistic as well as controlled observations (Rosenthal, 1962) and have devised ways and means to neutralize them (Carl Smith *et al.*, 1978). The exclusive goal of mainstream psychologists has been, and still is, to acquire knowledge that is reliable and ecologically valid so that such knowledge is effectively applied to correct behaviourally deviant and psychologically disturbed or deficient individuals, in human resource development as well as management, and in educational procedures, especially in the case of mentally challenged and learning disabled children.

Adhering to the principle of parsimony (Occum's razor) cognitive psychologists assume that mind is matter that functions as a machine in each individual. Under this assumption cognitivists have acquired reliable and ecologically valid knowledge about basic mental processes, such as consciousness, attention, perception, memory, representation of knowledge, decision-making and logical thinking. It has now become an integral part of a broad field of knowledge called cognitive science. It answers questions about the nature of knowledge, its components, development, and application. By virtue of its interdisciplinary approach cognitive science has brought cognitive psychology, neuroscience, computer science, linguistics and anthropology under one umbrella. Now empirical findings and their theoretical explanations are cross-examined, validated or rejected.

In summary it can be stated that in any area shrouded in pitch darkness knowledge is like a lighted spot. Anyone can observe what objects are there, and what events are taking place. Knowledge thus acquired is subject to scrutiny. Now the question is what is faith?

Distinctive Features of Faith

Faith is a mental product based on sensory inputs, perceptually encoded in terms of schemata, developed in accordance with ones upbringing, socialization, education, and supported by opinions of fellow humans and persons of authority. Subsequently one holds onto it tenaciously. It does not require objective proof. It functions as knowledge. Owing to confirmatory bias of human beings faith invents proof and confirms it through heuristic reasoning. A well-developed faith is hard to extinguish because it is *self-reinforcing*. However it is not scientific knowledge.

Faith-energized scholarship in psychology has endeavoured to counter the spread of scientific knowledge. It is pertinent to take an example. Soon after the formal beginning of scientific psychology in 1879 by Welhelm Wundt, a large number of faith-inspired scholars felt threatened. They met in London and accused psychology of the guilt of ignoring universally accepted genuineness of numerous supernatural psychic phenomena. In 1885 they established a 'Society for Psychical Research'. It is interesting to note that they had tenacious faith in the existence of psychic power and its universal manifestations. They too considered the empirical method as used by then psychologists, to be the most appropriate one. As a result widespread empirical studies of psychic powers, such as telepathy, clairvoyance, extra-sensory perception, reincarnation, dowsing, haunted house etc. were started. In many countries the aforesaid kind of organizations were established. Their empirical studies by the researchers were published mainly in two journals, *i.e. 'Journal of Parapsychology'* and *'Journal of the American Society of Psychical Research'*.

During this very period two books, one by Myers (1903), entitled *'Human Personality and Its Survival of Bodily Death'* and another one by J.B. Rhine (1935) entitled *'Extra-Sensory Perception'* were published and became widely read works. During the first seven-eight decades of the twentieth century parapsychological researches were continuously conducted in more than fifty university laboratories and medical colleges. Some Indian Universities also introduced parapsychological courses in psychology departments or established independent departments of parapsychology. This movement has failed to prove the genuineness of a single psychic phenomenon. Now this movement has dwindled to the status of insignificance, partly because of its failure, and largely because of fast increasing scientific knowledge and medical advances by virtue of which naive faith in existence and power of evil sprits, ghosts, demons, malevolent defies and magical powers of spiritual healers, has almost disappeared from the common man's world except in some scheduled tribes and illiterate people who are still somewhat enwrapped in those faith reinforced superstitions.

In the last five decades another marginal trend in contravention of the mainstream psychology has emerged. It is called 'cross-cultural psychology'. It has been initiated and supported by a select group of psychologists from all over the world. A perusal of their research writings makes it evident that they are against the unity of psychology as a science, and some of them are bent upon justifying their faith in cultural pluralism while adopting ostrich approach to cultural universals and unduly stressing cultural particulars. They also intentionally or unwittingly ignore the perilous problems of cultural dichotomies in a globalizing society (Hermans & Kempen, 1998). It has now become clear that these psychologists share four tenets upon which cross-cultural psychology is being built. These tenets are: (1) culture out there as context, (2) cultural pluralism, (3) cultural specificity of mental processes and human behaviour, and (4) urgent need for fractionation of psychology into multiple indigenous psychologies.

It is more interesting to note that the aforesaid tenets have not been accepted as assumptions for derivation of hypothesis so that their empirical verifiability/falsifiability could be assessed in an objective manner. Most of the cross-cultural psychologists have endeavoured to confirm them as tenets with the help of correlational researchers (selective manipulation of culture as the context variable) yielding quantitative differences among selected subjects. Use of this kind of designs by cross-cultural psychologists reflects their closed mindedness instead of their search for understanding. A curious reader expects specific referents of the construct of culture and clear cut meaning. Misra (2002) is of the view that the construct of culture is rooted in variformities, and it varies with the purpose of the user, and its definition depends on proposition or theory about those aspects of the world that can fit in any meaning that suits one's purpose of justifying one's faith in cultural pluralism. Any parameter that supports such implicit faith is accepted without examining it critically.

A number of cross-cultural psychologists experience discomfort when they are near the contemporary main stream psychology. They give two reasons for this feeling. First reason they articulate is that such a psychology is mechanistic, individualistic and reductionistic. However, they do not spell out what it should be, and in what respect psychology would be better and more effective as a scientific endeavour. Should psychology be animistic, collectivistic and constructinistic? The other reason, they give is that contemporary psychology is mono-cultural practice of Euro-American psychology. Such an explanation of their discomfort is odd. Do all European and American nations have the same culture ? It is a well known fact that these two continents are inhabited by numerous ethnic groups, have diverse religious rites and rituals, historical background, languages and diverse eco-cultural context. It seems reasonable to infer that such psychologists have certain faith about the nature of human beings, and they are aiming at instilling their faith in students of psychology in India.

One must remember that atheism and materialistic approach to understanding different facets of life in this world is not of Euro-American origin only. Long time ago it originated in India in the form of Charvak's philosophy which was further refined and elaborated by Charvak's disciples. Democritus in Greece was one of the early atheist philosophers who developed materialistic account of the world in Greece. From this vantage point it is reasonable to state that neither the West nor the East should be lauded or blamed for practicing or adopting the mechanistic approach to its knowledge seeking endeavours. As a matter of facts it is worth noting that Charvak enunciated his approach in the same historical era in which it originated in Greece. However, it is more relevant to remember that such a philosophy developed as a revolt against the Brahmanical hegemony of Upnishadic philosophy and before that too the emergence of the Jain and Buddhistic philosophies. Charvak's philosophy took birth and developed against the high sectarianism of Upnishads and Vedic rites and misuse of rituals as means of livelihood by the Brahmins and also against the then prevailing custom of sacrificing animals (at the altars of deities) in 'Yajya' and many other symptoms of disorder and instability rampant in social and potential life etc. (C.D. Sharma, 1960, p. 21). Subsequent development of Jain and Buddhistic philosophies also nurtured atheistic view point.

It is indeed amazing that some cross-cultural psychologists consider machines and mechanistic approach as product of the West. They seem to be wrapped up in the nineteenth century conceptualization of machines and are oblivious of the mechanistic approach adopted in Greece and India. Old religious scriptures of India give vivid description of automatic machines, air planes (*viman*) and a variety of weapons created for destructive purposes. Thus machines and mechanistic approach are neither western nor oriental in origin. The only difference between the two is that the western machines are real, and function because of being driven by physical force, while Indian machines mentioned in different types of scriptures were creation of imaginative faith in supernatural, psychic or spiritual forces. In the later half of the twentieth century conceptualization of machines has undergone sea change. Kurtzman (1987) borrowing heavily from Ashby (1956) and Haugeland (1978) writes that 'Cognitivism adopts the cybernetic conception of machine as its model'. A machine is an entity that can occupy one or more states. A state is understood informally as any well defined condition or property that can be recognized if it occurs again. These states transitions are determined by input to the machine from some external source, the machine emits output. He further states that three features are crucially characteristic of the machine that minds, both natural and artificial ones (intelligent computers). Computer simulation studies have gone a long way in unravelling numerous complexities of mental processes, such as, visual and language processing, problem solving, concept learning etc. Understanding enhanced by such studies has not only extended the boundary of knowledge, but has also increased applicational capability. Faith alone stands in the way of accepting this point of view by some specific brands of psychologists.

Indigenous Psychologies

Fractionation of universal psychology into multiple indigenous psychologies is one of the most important agendas of some cross-cultural psychologists. The notion of universal psychology is unacceptable to them because of its Euro-American origin. They think that its acceptance in India is just a manifestation of colonial past (D. Sinha, 1986). A number of other cross-cultural and Indian psychologists agree whole-heartedly with this view point, and exhort psychologists working in India to study and teach rich intellectual tradition and philosophical treatises of psychological significance on yoga, writings of various sects of Buddhism containing psychological concepts, methods of investigation, and techniques of self-control. However, one of the most interesting and thought provoking suggestions is that the intellectual traditions of India have potential universal applicability and must be exploited. At this point cultural pluralism does not stand in the way of universal applicability of indigenous Indian techniques of self-control.

Some promoters of indigenous psychologies refer to yoga as the technique for attainment of self-control. It is rooted in ancient culture of India. Such a technique is called *'ashtang yoga'* and is an applicational extension of Samkhya philosophy which posits two basic realities, *i.e.* soma and psyche (*Prakriti* and *Purush*). The technique of yoga is prescribed for those who are interested in maintaining good physical and mental health, and are also interested in getting liberated from the travails and woes of worldly life and in attaining the status of pure and blissful consciousness. This yoga consists of eight components or layers. The first two, *yam* and *niyam* are general rules of conduct to be followed in everyday life. They lay emphasis on non-violent and other-oriented behaviours. The next two components are concerned with physical exercises called *asan* and *pranayam*. These exercises lead to healthy heart and body. The fifth component, called *pratyahar*, prescribes for voluntary control over the natural tendencies of receptors and effectors. One has to resist extroversion and to be attentive to one's own bodily and mental activities. Thus, these five are regarded as external means of control, and can be empirically evidenced.

The sixth and seventh components are called *dharana* and *dhyan*. They aim at controlling attentional processes of consciousness and adapting them to acquiring capability for concentration. It means that doing these two exercises one becomes completely adept in investing one's entire attentional capacity for fixation on one icon or image preferably some idol. After attaining these two stages of consciousness, one becomes thoroughly self-controlled and self-regulated. Such a self-controlled person becomes mentally as well as physically healthy.

Upto this point it is relevant to note that indigenous Indian psychological insight extracted from yoga is in complete harmony with the mainstream psychological techniques and methods for acquisition of self-control. The mainstream psychologists have experimentally demonstrated that involuntary responses, such as, visceral and glandular ones, can be brought under self-control. Now, it has become a well acknowledged fact that self-control can be learnt using instrumental control procedure (Miller, 1969). Experimental studies have demonstrated that voluntary control of striped muscles leads to development of skills for the control of ones own brain activities (Kamia, 1969; Brown, 1970; Green *et al.*, 1970). Experimental studies have convincingly demonstrated that self-control and self-regulation are acquired through perceptual learning (Murphy & Kovach, 1977). Carver and Scheier (1981) authored a book entitled *'Attention and Self Regulation: A Control Theory Approach To Human Behavior'*. They have presented an elaborate account of self-control, self-regulation and how perceptual learning and attentional focus lead to self-efficacy culminating in self-control and self-regulation. In this respect the East and the West have met, and have developed different products which fulfil the same needs. The only difference between the two is that they use different techniques, technical terms and meta-theoretic assumptions.

The eighth component of yoga is called *samadhi*, some extraordinary type of trance. This topmost layer of yoga passes through a variety of developmental stages. Ultimately it culminates into an altered state of consciousness in which the practitioner attains the purest and eternal consciousness which assumes the form of pristine psyche (*Purush*) completely liberated and detached. The lures and stratagem of *'Prakriti'* or pristine nature are now utterly ineffective on that form. In this state of pure consciousness the attender, the attended and the attentional process get merged into each other and become undifferentiable as a unity. It is an altered state of consciousness that is totally emancipated from the worldly sufferings, from compulsion of actions for survival, and also from cycle of birth and death.

At this point a rational question arises: once an altered state of supreme consciousness is accomplished in which the three elements—attention, attender and attended are united into one entity of pure consciousness, how come that it again gets differentiated in earlier forms? What kind of volition is there to redifferentiate the three from this supreme consciousness? Such a rational question has not been answered. One has to accept the reality of such an altered state of consciousness by having faith in yoga guru's (teacher's) knowledgeability and integrity. However at this stage, this altered state of consciousness becomes an indescribable mystery. A knowledge seeking person, not prone to develop blind faith is resistant to accept its reality. In such conditions mystery thickens. The practitioners and promoters of yoga exhort and persuade others with the help of their emotional intelligence to practice yoga and realize that kind of supreme consciousness. A rationalist may not be persuaded to comply, but the credulous ones do. It is for such credulous masses that different brands of yoga such as *shabd, hath, zen, transcendental* meditation etc. are available in the world market.

It seems reasonable to look at this altered state of consciousness from another vantage point. It is widely known that there are a variety of trance states that create conditions of complete mental absorption or deep musing and elation. Such trances are not occult phenomena. A variety of neurophysiological researches have shown that consciousness reflects processes that occur in several areas of brain or among several neural circuits which are represented in areas of brain (Dennet, 1991). Brain lesion studies and the techniques of regional cerebral blood flow, evoked potential and single cell recording have gone a long way in deciphering numerous complexities of consciousness and a variety of cognitive functions.

Apart from the fact that some altered states of consciousness, such as dissociative hysteria, fugue, dual personality and somnambulism etc. are malfunctional, but dreams are universal phenomena. The techniques of bio-feed-back and brain-imaging are used with a view to deciphering the unknown aspects of cognitive processing in various altered states of consciousness. Numerous altered states of consciousness are induced by drugs. Tranquilisers and antidepressant drugs are universally used for restoring consciousness to its normal functioning. On the other hand use of psychedelic drugs are not of recent origin. Chemical derivatives, now called cilocybin and mascaline, were used by preliterate humans of Venezuela and Columbia. These people used them in crude form for experiencing ecstasy. These drugs were claimed to be mind-revealing resulting in religious revelations. Psychedelic substances, such as *cannabis indica*, marijuana, opium and *som-ras* (as mentioned in religious scriptures) were commonly used in India. In modern times use of psychedelic drugs including opium, toddy (the sap of palm trees), *cannabis indica*, marijuana, lycergic acids, brown sugar, LSD etc. has become a serious problem at international level, and also a big business. These drugs are consumed or used for experience of indescribable ecstasy. Systematic medical studies of the effects of such drugs have shown that they lead to multistage effects ranging from simple sensory effects like enhanced visual vividness to loss of meaning, transformation of one's basic identity to depersonalisation and religious ecstasy. The question of whether psychedelic drugs produce experience distinguishable from profound, spontaneous, meditational mystic experiences has been subjected to a variety of neurophysiological studies. Pahnke (1966) has demonstrated that LSD induced states of consciousness are not distinguishable from the non-drug states of the higher mystical levels of consciousness. Personal faith in theism would lead to immediate rejection of the aforesaid findings as utterly heretical. However, a self-controlled, neutral and detached stance regarding the reality of the final stage of *samadhi* in which pure and pristine form of consciousness is attained is not followed.

Distinctive Features of Indigenous Indian Psychology

Let us focus attention on another aspect of development that is going to further befudge the line of distinction between faith and knowledge. This development aims at creating an indigenous Indian psychology. Numerous statements made by promoters of the Indian brand of psychology make it apparent that the newly designed indigenous Indian psychology is going to be completely different from the contemporary mainstream psychology alleged to be of Euro-American origin. It is proposed to be a humanistic discipline settled in Arts faculty, theistic and transpersonal in its meta-theory, multimethod, multiple interpretional framework and highly elastic and malleable in conceptualisation. One gathers an impression from their writings that its applicational extension is going to be physical, mental and spiritual health as well as well-being. It also becomes clear that their emphasis on library research and arm-chair speculation will overweigh empirical research. For them Indian philosophical treatises are gold mine of psychological insights. Use of scientific methods are going to be abandoned because they are of Euro-American origin. Their acceptance is reflection of colonial mentality.

The idea of developing an Indian psychology is not new. The only novelty is to add another adjective—indigenous to it. During the hay days of the schools of psychology in the West, the idea of having another school or system in India was raised. Numerous text books of Indian psychology were published and included in the curriculums of psychology in a number of Indian universities. However, it is interesting to note that as the schools of psychology disappeared and their substantial components got integrated in mainstream psychology, so did Indian psychology except from a few psychology departments.

It is surprising that the reason for creating indigenous Indian psychology is not rooted in demand for more precise and well-honed knowledge, but in faith generated sentiments against the mainstream psychology. The reasons advanced by some advocates of indigenous psychologies are varied, such as, (1) uncritical acceptance of empirical methods may hide an underlying cultural imperialism, a legacy of the colonial past, (2) need for promotion of multiple indigenous perspectives and for weakening univocalist demand, (3) techniques of experimentation and instruments have syncretic limitations for studying transpersonal phenomena. These reasons need no comments. It is enough that numerous philosophical cults preach for acquisition of self-control and detachment from world by humdrum and involvement, argue in favour of constructionistic approach, but find neutral, self-controlled objective detached observation as a cult.

It is most incredible that indigenous psychologists argue and plead in favour of reviving the percepts and principles of ancient Indian philosophies. In doing so they evince self-serving bias and closed mindedness in respect of psychohistorical facts. Indian history is characterized by recurrent intellectual protests against theistic cults and exploitative brahminical hegemony. However, it is indeed amazing that theism and its pluralism has evinced tremendous resilience and kept the country divided into numerous small sized kingdoms, castes and sects. It is no wonder that this country remained vulnerable to alien forces and finally, was enslaved by the Britishers.

Quite often in informal discussions and exchange of ideas it becomes transparent that the move for creating indigenous psychologies is a camouflaged political agenda of some intellectuals. However, in their thinking such intellectuals remain oblivious of an anomaly in their thinking. It is certain that the country has to develop plans and strategies for future growth and progress. It has to move forward and get rid of outdated and incapacitating hangovers. In this background, attempt at revival of ancient Indian perspectives and principle is a blatant regressive step.

Faith and Cultural Pluralism

In all academic disciplines—physical, biological or cognitive sciences and humanities, all wordly objects, events and entities are placed in dual categories or located on bipolar dimensions. Cross-cultural, cultural as well as indigenous psychologists locate or place culture on the dimensions of singular-plural, uniform-variform, particular-universal, and product-process. According to these psychologists culture as a social phenomenon is characterized by pluralism, variformity, particularity and product quality. Culture as a product is treated as behavioural context or medium. However, when these psychologists are required to define, delineate and exemplify its referents, confusion becomes confounded. According to Price-Williams (1989), meaning of culture hinges on where does it enter. D. Sinha (1996, 2002) states that 'one comes across great diversity in the way culture has been viewed in relation to behaviour. Even if we take the nomenclature, there are many diversities for beyond the level of fruitful exchange of ideas.' In this kind of scenario to consider culture as a product out there to be studied, and then load it with surplus meaning to suit one's arbitrarily chosen perspective creates an academic chaos.

At this point of conceptual confusion, it seems necessary to get rid of the current mental set, and adopt a neutral stance for defining and characterizing the construct of culture and its referents. There is no gainsaying the fact that culture is a human creation or product. All human products—material or symbolic—are constituted of four components. They are (1) raw physical or mental material, (2) ecological, social or personal demands, constraints and compulsions, (3) methods and expertise, and (4) potential force, motive or urge. It is important to note that for creation of any product these components are singly necessary and jointly sufficient. It is obvious that these components are easily separable from one another. The first three components are highly variant in relation to place and time of creation. It is important to note that the Zeitgeists vary from one historical era to the other, and ecological and social conditions vary from one geographical area to the other. Because of such variations human products also vary.

However, the most important point worth remembering is that the fourth component, *i.e.* the potential force or motive that underly all product remains constant. It is, thus, reasonable to assert that this potential force creates antecedent condition for a creator, to become active and lead to consequent condition of creating some product. What activates this potential force is inherent in adjustive—accommodative as well as manipulative—exploitative need of human beings in relation to the environment. Once activated this potential force manifests itself in cultural processes. These processes are guided by the principle of best efforts for making life more comfortable, pleasurable, intrinsically satisfying and expertise enhancing. As the human knowledge grows and the circle of enlightenment expands, cultural processes become innovative and the cultural forms are altered over time and across space. With increase in knowledge the darkness shrouded area of faith shrinks. As a result the faith guided individuals feel threatened and try their best to suppress knowledge. The cases of Gautam Buddha, Galileo, Aristotle and a host of others alike provide robust testimony of this fact. It is a widely known historical fact that in the United States of America the entry of science as subject to be taught in schools was blocked for many decades by sectarian thinkers, armed-chair speculators and philosophers. The tenet of cultural pluralism is reflective of a kind of the aforesaid faith.

All cultural processes are continuous variables and their manifestations undergo change with change in the first three components of a cultural product. That is precisely because of this that the exact number or kinds of culture has yet not been counted and clarified. Advocates of cultural pluralism ignore the boundary areas between two *prima facie* distinguishable cultures. Moreover it is interesting to note that even conspicuous cultural similarities are thrown in the limbo of oblivion. The problems of cultural dichotomies in a globalizing society (Hermans & Kemplen, 1998) are ignored. The propounders of cultural pluralism stand in contradistinction with the idea of seminal unity of mankind and intrinsic unity in the manifest diversity across space and over time.

If one makes an incisively critical analysis of the empirical referents of the construct of culture it becomes irrefutably clear that cultural universals are far more numerous, pronounced and impressive than cultural particulars. The frequently cited referents of culture are (1) knowledge, (2) language, (3) religion, (4) social structure, (5) art and architecture, (6) customs and tradition, (7) form of government, (8) livelihood activities, (9) metaphors and symbols, and (10) values. When casually observed from a distance these referents appear to be different on the whole. However, if one delves deep in their variformity, the substance contained in their forms, the inherent potential force behind them, and the processes as well as their purposes and processes, the cultural unity stands out.

Let us take the case of spoken as well as written languages. Each and every language appears to be distinctively unique in its phonic and visual features. However, when their structural and functional features along with their origin are analyzed, the inherent similarities among them stand out. Now it has become an irrefutable fact that each language originated in biological adaptation processes, and in no way can be considered as a cultural invention. It is now well known that there are linguistic areas in the human brain situated in and around Broca's and Wernickes areas (Guenther, 1998). The former functions to sequence the sounds of language, and latter functions to associate words to their meanings. It is now also known that all languages have structural similarities. Each one is constituted of phonemes, morphemes, syntax, semantics, grammars and deep and surface structures, and their paralinguistic and non-verbal supports reflect several universal features of languages (Hocket, 1996). It is precisely because of these universal features that one can learn one or more alien languages and may become effective bilingual or multi-lingual.

All religions, as referents of culture; irrespective of their Asian, European or Arabic origin, are essentially monotheistic in their faith. All religions were conceived and founded by highly inspired humans. Each religion has two distinct facets, *i.e.* extrinsic and intrinsic. Extrinsic facet of each religion consists of specifically defined modes of worship and prayer, ecologically suitable rites and rituals, well defined superstitions, and rules as well as mores of life-style. On this basis followers of every religion are easily distinguishable. Once upon a time rules for clothings were easily distinguishable. In the early times of each religion extrinsic facets were unexceptionally conformed to. Any visible norm violation by individual used to culminate in ostracization. However with fast occurring liberalization and changing Zeitgeist, practice of extrinsic facets is now limited to fast shrinking numbers of religious persons.

The intrinsic facet of all religions lays emphasis on conduct and character, ethical values, such as truthfulness, honesty, kindness, compassion, altruism, self-restraint, sense of detachment, simple living and minimization of cravings for hedonistic activities except those that are necessary for healthy survival and mental peace. In this respect all religions are maximally similar except in differential emphasis on ethical principles. It is because of the intrinsic religiosity that in numerous countries of the world, communities of different religions are used to amicably coexist and are moving forward on the path of making life more pleasurable and meaningful. The features of intrinsic religiosity do not fit in and is dissonant with the tenet of cultural pluralism. Extrinsic religiosity because of its variformity may create an illusion of cultural pluralism, but substantive processes inherent in them are also highly similar.

Knowledge as an empirical referent of culture is extracted from common man's lexicon in which veridical and tenacious knowledge becomes inseparable. In a book edited by Kroeber (1966) it has been made quite clear that knowledge implies two sets of cognitions, *i.e.* scientific knowledge and mythological knowledge (pp. 121-128). It is a historical fact that in ancient times myths and mythologies were created by the guardians of religions supported by theologists in the West and theistic philosophers and preachers with a view to strengthening their hold on the masses, and also to make their preachings and discourses more lucrative. On the other hand scientific knowledge is relatively of late origin and is universally applicable and valid. In modern times the number of scientific knowledge guided humans in all developed and developing nations is fast increasing resulting in shrinkage of faith guided members of societies. Such a fact puts a question mark against the tenet of cultural pluralism.

If one critically analyzes the other empirical referents of culture, it becomes crystal clear that culture is not an objectively identifiable behavioural context. Culture, in reality, is a set of mental processes that occur in human activities. These processes are operative in individuals and should be studied in the same way as other basic cognitive processes and be subjected to empirical studies in cognitive science.

In the end it is necessary to accept faith as an essential ingredient for making one's personal and private life more peaceful and useful. It is so because the knowledge illuminated area in the infinitely vast space of ignorance is still very limited. However faith becomes obnoxious when efforts are made to impose it on others. Such efforts are exploitative and stand against, and corrode, the principles of intrinsic religiosity.

References

Ashby, W.R. (1956). *An Introduction to Cybernetics*. London: Chapman & Hall.

Brown, B.B. (1970). Recognition of aspects of consciousness through association with E.E.G. alpha activity represented by a light signal. *Psychophysiology, 6*, 442-452.

Carlsmith, J. Merril, Ellsworth, Foebe, C., & Aronson, Elliot (1976). *Methods of Research in Social Psychology*. Massachusetts: Eddison-Wesley.

Carver, Charless, & Scheier, M.G. (1981). *Attention and Self-regulation: A Control Theory Approach to Human Behaviour*. New York: Springer-Verlag.

Dennet, L. (1991). *Consciousness Explained*. Boston: Little Brown.

Festinger, L. (1954). A theory of social comparison processes. *Human Relations, 7*, 117-140.

Freud, S. (1922). *Beyond the Pleasure Principle*. The Standard Edition (Vol. 18) London: Hogarth Press.

Gamst, Frederick, G., & Norbeck, Edward (1976). *Idea of Culture: Sources and Uses*. New York: Holt, Rhinehart & Winston.

Gardner, H. (1985). *The Minds New Science: A History of Cognitive Revolution*. New York: Basic Books.

Green, E.E., Green A.M., & Walters, E.D. (1970). Voluntary control of internal states. *Psychological and Physiological Journal of Transpersona Psychology*, 2, 1-26.

Guenther, R. Kim (1998). *Human Cognition*. NJ: Prentice Hall.

Hermans, Robert, J.M., & Kempen, J.G. (1998). Moving cultures. *American Psychologist, 55*, 1111-1120.

Haugeland, J. (1978). The nature and plausibility of cognitivism. *Behavioral and Brain Sciences*, 2, 215-226.

Heider, F. (1958). *The Psychology of Interpersonal Relations*. New York: Wiley.

Hocket, C.F. (1966). *Universals of Language*. Cambridge: M.I.T. Press.

Kamiya, J. (1969). Operant control of the EEG Alpha rhythms and some of its reported effects of consciousness. In C.T. Test (Ed.) *Altered States of Consciousness*. New York: Wiley.

Kelley, H.H. (1967). Attribution theory in social psychology. In D. Levine (Ed.) *Nebraska Symposium on Motivation*. Lincoln: Nebraska Press.

Kurtzman, Howard, S. (1987). Deconstruction and psychology: An introduction. *New Ideas in Psychology, 5*, 33-71.

Marx, Melvin, H., Cronan-Hillex, and William, A. (1987). *Systems and Theories in Psychology* (new ed.), New York: McGraw Hill.

Maslow, A. H. (1943). A theory of human motivation. *Psychological Review, 50*, 370-396.

Miller, N. (1969). Learning of visceral and glandular responses. *Science, 163*, 434-4453

Misra, G. (2002). Introduction. *Journal of Psychology and Developing Societies, 14*, 1-9.

Murphy, Gardner, & Kovach, Joseph, K. (1977). *Historical Introduction to Modern Psychology* (Sixth ed.). London: Routledge & Kegan Paul.

Olds, J., & Milner, P. (1954). Positive reinforcement produced by electrical reinforcement of septal areas and other regions of rat brain. *Journal of Comparative and Physiological Psychology*, *47*, 419.

Pahnke, W. (1966). Drugs and mysticism. *International Journal of Parapsychology*, *8*, 295-320.

Rhine, J. B. (1935). *Extra Sensory Perception*. Boston: Humphries.

Rosenthal, Robert (1966). *Experimenter Effects in Behavioral Research*. NewYork: Appleton-Century-Crofts.

Sapiro, Melford E. (1978). *Culture and Human Nature in the Making of Psychological Anthropology* (ed). Berkeley and Los Angeles: University of California Press.

Sharma, C.D. (1960). *A Critical Survey of Indian Philosophy*. London: Hider & Co.

Sinha, D. (2002). Culture and psychology: Perspective of cross-cultural psychology. *Psychology and Developing Societies*, *14*, 12-25.

6

Some Methodological Problems of Deprivation Studies

Over the past few decades there has been phenomenal increase in the quantum of psychological research on the behavioural and cognitive impacts of deprivation. This upsurge has yielded data leading to contradictory results and mushroom growth of theories. Inconsistencies in results and confused theoretical speculations may be primarily due to a variety of methodological problems and shortcomings with which the contemporary deprivational studies are fraught. Scientific endeavours to look into the relationship between deprivation and the emergence, growth and level of psychological functions are of recent origin. Therefore, neither has the concept of deprivation become sharpened and crystallize enough to include in its orbit all its phenomenal or empirical referents in order to be accepted as a respectable scientific concept, nor has the methodology employed in measuring deprivation, as well as in designing deprivation studies, become sophisticated enough to yield unequivocally valid and reliable data. In the present paper an attempt has been made at identifying those methodological problems and shortcomings which have to be tackled in order to make deprivation studies scientifically sound and functionally meaningful.

Research efforts in the area of the psychological effects of deprivation have proceeded in two distinct directions. On the one hand, deprivation has acquired the status of an operational concept and has been treated as an experimental variable in controlled laboratory conditions. On the other hand, it has also acquired the status of a natural phenomenon. It is manipulated through some selection procedure and is usually treated as a nominal, or at best, as an ordinal variable (Whiteman & Deutsch, 1968). The laboratory treatment of deprivation has given rise to the term of sensory deprivation in which so widely divergent operations are used that the popularly known term of sensory deprivation has become a misleading term. The concept of deprivation in natural settings has also been so freely interpreted that for a student of psychology it has now become almost impossible to understand its exact meaning. In both types of researches the numerosity of methodological problems and issues has been on the increase as the research output has increased. Although these two lines of distinctly separable research developments may be viewed as having a common ancestry, and proceeding towards a common future, they have come to be beset by different sets of methodological problems.

Broadly speaking, laboratory studies of deprivational effects are truly experimental while its studies in natural settings are largely correlational. From this obvious fact it can easily be surmised that the methodological problems, issues and considerations confronting the investigators in these two types of researches must be of different complexions and complexities. It is, therefore, reasonable to treat them under separate headings. It is true that sources of the emergence of many problems in these two types of investigations are more or less similar but they assume different characteristics in their functional forms. Since experimental investigations into the effects of deprivations on psychological functions have a longer history and have attained greater precision than the deprivation-research in natural settings, their methodological problems have become more crystallized (Rossi, 1969). The present analysis shall focus on behavioural impacts of deprivation as it occurs in the natural life setting.

During recent years emphasis on studies of deprivation has shifted from laboratory to deprivation in real life conditions of individuals, and numerous research papers on the effects of deprivation on a host of psychological functions have been published. However, a critical survey of these studies reveals two sets of methodological problems. The first set of problems originates from diversities in the meanings of the term deprivation and the criteria used in its identification as well as its measurement. The second set of problems originates in actual conduct of such deprivation studies. The former is primarily an issue of devising ways and means for attaining precision in the measurement of the nature and extent of deprivation, while the latter set of problems is concerned with the adequacy in control of the confounding variables and eliciting reliable and valid responses to various objectively scored test items.

Conceptual Diversity and Measurement Problems

In the studies of deprivational effects in natural settings psychologists and other behavioural scientists have conceptualized the term 'deprivation', not as scientists, but as litterateurs imposing subjectively determined connotations to the term and selecting arbitrary criteria for separating the deprived from the non-deprived. More often than not, knowingly or unknowingly, psychologists have become much more motivated by expediency and convenience rather than by methodological soundness and exactitude in determining the empirical referents of deprivation. As a consequence of this, a multitude of terms have emerged each purporting to signify deprivation in this or that sense leading to diversity in measurement criteria and devices. It would not be out of place to present a flavour of the terminological diversity by sampling some of the terms that have been used as synonyms of deprivation. This terminological sample will help us in understanding the difficulties that are likely to arise devising ways and means for quantifying and assessing the nature and extent of deprivation to which an individual has been subjected in his life. Some of the terms used are: *effective deprivation* (Gewirtz, 1961); *cognitive deprivation* (Green, Hofmann & Morgan, 1967); *cultural deprivation* (Riessman, 1962; Kogan & Kogan, 1970; Wight, Gloniger & Kneene, 1970; Das, 1973); *economic deprivation* (Symmonds, 1968); *educational deprivation* (Getzels, 1969); *emotional deprivation* (Stott, 1974); *environmental deprivation* (Deutsch, 1965); *inequality* (Miller, 1968); *parental or maternal deprivation* (Yarrow, 1961); *malnutrition* (Das & Pivato, 1977; Dobbing, 1967; Bladslee, 1967); *medical deprivation* (Suchman, 1966); *poverty* (Bee, 1974); *prolonged deprivation*

(Misra & Tripathi, 1977a; Tripathi & Misra, 1975, 1976); *psychological deprivation* (Langmeièr, 1972; Goldfarb, 1945; Langmeier & Matezeck, 1975); *relative deprivation* (Crosby, 1976); *social pathology* (Coleman, 1971); *social and cultural disadvantage* (Havighurst, 1964; Gordon, 1968; Uzgiris, 1968; Sinha, 1977a; Singh, 1976); *social deprivation* (Rankin, 1970; Tulkin, 1972).

All these terms have inevitably led to different criteria for isolating the empirical referents of deprivation. These empirical referents differ from one author to the other, and at times the confusion is so pervasive that what is conceived to be deprivation in one approach is not necessarily considered to be so in the other.

Most of the investigators have used the concept of deprivation, irrespective of the difference in deprivational magnitude, as a nominal or categorical variable. It has been tacitly assumed that deprivation is a characteristic which is either present in an individual's life, or it is non-existent. They classify their universe of study in two groups, *i.e., deprived* and *non-deprived*. For such a classification of people they adopt a single criterion of this or that kind, and on the basis of that criterion they separate the deprived from the non-deprived. Let us examine some of these criteria and evaluate their adequacy in measuring deprivation.

Residential Area

A number of investigators (*e.g.*, Wight, *et al.*, 1970; Gordon, 1968; Adams, 1971; De Board, 1968) have employed residential area as the sole criterion for distinguishing deprived individuals from the non-deprived ones. It cannot be denied that living in certain areas may have some degree of correlation with deprivation but it cannot be considered as the valid basis for identifying the 'deprived'. More often than not, location of one's residence is a matter of heritage and in a relatively non-mobile society, such as one in India, some residences, even in the worst areas of a township or a village, have all the facilities necessary to keep materialistic deprivation away from life, and at times presenting a contrast to the adjoining residence creates a sense of enrichment in its inhabitants leading to perhaps significant psychological impacts.

Institutionalization

Recently a number of studies have been reported in which institutionalization of children has been characterised as deprivational. Observed differences between the institutionalized children and those living in their own families have been ascribed to deprivation. Orphanages (Sinha & Shukla, 1974), foster homes (Talmadge, 1969) and child care centres (Langmeier & Matezeck, 1968) are some of the institutions identified as deprivational. Undoubtedly such institutions present deprivational setting to a great extent but it cannot be said that they are always deprivational. However, deprivation can occur and does occur in ordinary family life also. The best that can be said about this criterion is that such institutions and normal families create different types of life conditions and provide not only different types of experiential deprivation but patterns also. There are a host of other variables also along which the two settings differ, and therefore, it does not seem sound and scientifically justifiable to consider them distinguishable in terms of deprivation alone.

Membership of Some Cultural/Social/Economic Group

One of the most frequently employed criteria of deprivation is membership of some specified group. Individuals have been considered as deprived on the grounds of low socio-economic status (Gordon, 1968), being Negro (Wilcox, 1971; Green & William, 1965), belonging to a minority community (Goff, 1954) or to an aboriginal tribe (Nurcombe, 1970) or to a particular caste (Das, Jachuk & Panda, 1970; Rath & Dash, 1972 ; Sinha, 1977). It cannot be denied that certain segments of the social structure in any country are correlated with deprivational life-experiences but it does not require much effort to discern that in modern socio-political atmosphere membership of specific groups or communities does not present insurmountable barriers for enriching oneself and becoming non-deprived. Every community today has a fairly sizeable number of families which provide enriching conditions of life-experiences. Moreover, deprivation cannot be considered as of a fixed category but as a dimension or a set of several dimensions along which quantitative differences may exist. Such categorical criteria fail to provide quantitative discrimination among the people of the so-called deprived categories.

Having realized the inadequacy of treating deprivation as a nominal variable investigators have started using a variety of other criteria for separating the deprived from the non-deprived with varying magnitudes and use indices of deprivation. This may be considered as the first set of serious attempts at quantification of deprivation. Davis (1968) used a set of rating scales for assessment of deprivational background, even though these rating scales covered only a very narrow range of background factors. A more refined attempt was made by Munsey (1971) in which he developed a Cultural Deprivation Index independent of ethnic group membership and included education, income, and area of residence. Whiteman and Deutsch (1968) developed another index in which a number of basic societal conditions were included. They argued that 'the association between social grouping and specific environmental factors is not directly causal but is mediated by some more basic societal conditions, such as unemployment, poverty and inequality of opportunity in various areas. With the removal of such conditions the association between social grouping and disadvantaging factors may vanish' (Whiteman & Deutsch, 1968; p. 91). Thus, they consider social deprivation as a relative term. They included six variables so combined in a measuring device as to yield a composite score known as "deprivation index". The index has been used by them in the roles of specifier, mediator, independent contributor and also as an interactive variable. The six criterion variables used in the index included housing condition, educational aspirational level for the child, number of children under 18 years of age in the family, number of cultural experiences, dinner time conversation and attendance of the child in kindergarten. These variables were included because they related to, (i) an important psychological function, such as reading, and (ii) social groups with varying socio-economic status. The development of the scale was made on a meagre sample of 294 children and no attempt at its cross-validation was made.

On similar lines Rath and Samant (1975) developed a cultural deprivation index suitable for Indian setting. Having derived some concepts and techniques from pertinent literature they selected 14 factors of socio-cultural disadvantage or deprivation. Following the method of equal appearing intervals, 42 categories, referring to the selected 14 factors were prepared, and 90 judges (postgraduate students) gave their ratings. Scale values derived from these ratings were determined for each category. The deprivation score of an individual was equal to the mean scale values obtained for all the 14 factors. A later study indicated that family income, father's education, place of residence, and caste of subjects were significantly related to performance on Raven's Progressive Matrices and verbal development. Undoubtedly this scale is an improvement over the other scales, and has included experiential areas in which deprivation may be considered critical for psychological functions. However, it is also open to criticism on more than one ground. For instance, this index was derived on the basis of the responses of judges. While this is an accepted procedure in attitude scaling it cannot be recommended for measurement of deprivation, for the reason that deprivation as conceptualized and understood by judges cannot be considered identical with that which occurs in real-life settings. Judges, being well-to-do and highly educated persons, cannot be considered as having first hand experience of deprivational existence. Their ideas of deprivation are mostly based on secondary sources of information and anecdotes mixed with various kinds of imageries. Therefore, it is essential that in measurement of deprivation the reference point must be the deprived group and not the judges. Items should be selected on the basis of close scrutiny of the living conditions prevailing in the environment of deprived individuals. Moreover, there are a number of important experiential areas that have not been included in this scale.

Although indexing of deprivation is decidedly an improvement over earlier categorical treatment of deprivation, there are still many shortcomings in this approach. In brief, five methodological shortcomings can be mentioned. Primarily, in indexing of deprivation the coverage of criterion variables is partial and narrow, and, therefore, limits the conceptual status of the term deprivation. Secondly, the scoring procedures have not been fully objectified. Elaborate administration and scoring procedures have not been standardized as a result of which investigator's subjectivity still affects the indexing process. Thirdly, the approach can be best characterized as treating deprivation as a nominal variable on the basis of a number of criteria instead of one. As a consequence, varying magnitudes of deprivation cannot be discriminated on the basis of such indices. Fourthly, the current indices of deprivation have left out the experienced or felt or phenomenological aspects of deprivation which have to be considered important in shaping human behaviour. Finally, the indexing of deprivation has laid emphasis on the prevalent status of the individual on the criterion factors and has not paid attention to the temporal aspects of deprivation. It can easily be imagined that any individual, no matter of what chronological age, must have had deprivational experiences not at the time of deprivational indexing but lasting over varying periods of time. Thus, deprivation can be conceptualized as having a temporal dimension which has been neglected in the current efforts at measuring it. It is interesting that in experimental studies of sensory deprivation while its temporal aspect has loomed very large, studies of deprivation in natural settings have ignored it. This shortcoming of the available tools has created a big barrier in attaining precision and comprehensiveness in deprivation measures.

Having realized the methodological limitations inherent in the above mentioned approaches to the measurement of deprivation, Misra and Tripathi (1977a, b, 1978) have developed Prolonged Deprivation Scale (PDS) in which deprivation has been taken as a more refined, comprehensive and experience-oriented concept. They have also laid the due emphasis on the temporal dimension of deprivation.

PDS made an attempt at identifying the various aspects of environment in relation to which individuals become enriched or impoverished (deprived). This scale included fifteen aspects of natural settings with special reference to life experiences in this country. These aspects comprise (1) housing condition, (2) home-environment, (3) economic-sufficiency, (4) food, (5) clothing, (6) educational experiences, (7) childhood experiences, (8) rearing experiences, (9) parental characteristics, (10) interaction with parents, (11) motivational experiences, (12) emotional experiences, (13) religious experiences, (14) travel and recreation, and (15) miscellaneous socio-cultural experiences. These areas have been split into 96 items along which the investigator is required to rate, on five-point scales, the information culled from the respondent on the basis of interview and observation. The scale is so designed that it can be used with maximum profit, if used at the residence of the respondent. The scale has been standardized on a large sample (n = 471) belonging to the age group of 15 to 25 years. This sample was drawn from all kinds of villages, larger and small, developed and underdeveloped, multi-caste dominated and single caste dominated, near townships or away from them. After standardization, data from 600 more respondents were collected and entire data were subjected to factor analysis. Two factors have emerged which have been designed as physical and psychological. The scale has been validated in variety of ways.

It can be unhesitatingly asserted that PDS presents a marked improvement over other scales and can be profitably used in all kinds of societies provided that its items are translated into the language of the locality in which research has to be conducted, and the investigators are well versed with the local dialect, customs, mores, and have the capacity to establish rapport with the respondent. However, there is still room for improvements in it. Even though the authors have tried to be as exhaustive as possible, it cannot be claimed that all possible deprivational areas of life have been included in it. The scale is based on the assumption that deprivation is a global phenomenon and has been so treated. Naturally, on the basis of this scale subjects with different types of deprivations cannot be discriminated.

On an *a priori* basis, it can be asserted that each area of deprivation might have differential effects on psychological functioning. But on the basis of this measure such hypotheses cannot be tested. Another limitation that can be pointed out is concerned with the conceptual framework of deprivation within which PDS has been developed. The scale treats lack of physical facilities as well as impoverished experiences as something absolute and does not give due consideration to deprivation as a relative phenomenon.

Macroscopic v. *Microscopic Levels of Measurement*

In literature on the effects of deprivation on cognitive growth one comes across a rather convenient and expedient method of separating the deprived from the non-deprived persons. Some nominal variable, such as race or caste or religion or residential locality or institutionalized-noninstitutionalized upbringing during childhood, is taken as the sole criterion on the basis of which persons are placed in either deprived or non-deprived category. This is justified as measurement of deprivation at its macroscopic level. This technique of measurement was conceived and conventionalized in the United States of America where two easily distinguishable races, *i.e.*, White and Black, exist. The Blacks have been labelled as deprived while the Whites as the non-deprived people. Faithfully following this technique, a number of psychologists in India have considered the members of scheduled castes or scheduled tribes as constituting the universe of deprivation and other castes as that of non-deprivation or enrichment. However, the subtle but significant differences between the two societal structures, *i.e.*, Indian and American, have escaped their notice. The distinction between the Whites and the Blacks in the United States of America is racial, has historical roots separating them economically, culturally, politically and more often than not segregating the Blacks physically as well as educationally. On the whole, the American society is dichotomous having two distinct social ethos, one with the traditions of "had-been masters", attitudes of superiority and with the rights, privileges and conveniences of "had-been owners", while, the other one with the deeply etched scars of "had-been-slaves", experiences of inferiority and with the liabilities, hardships and inconveniences of the "had-been-owned". Thus, in that country if the two ethos are considered separately, one consisting of the deprived and the other of the non-deprived, there does not appear to be any methodological snag, though on other methodological grounds of precision and soundness, the practice can be subjected to valid criticism.

However, this kind of generalization or description cannot be validly transferred in case of the social structure in India. The Indian society can be more aptly described as hierarchical. It is true that, usually, the lower the hierarchial status of a caste, the greater is the probability of its being impoverished economically, culturally and educationally. But this relationship is not absolute. Therefore, the dichotomization of Indian subjects on the basis of caste or religion does not have the merits of the White-Black dichotomization.

Apart from this shortcoming on the grounds of differences in the two societies, there is another danger which is inherent in the macroscopic methodological approach. It is self-evident that when one talks of macroscopic method of measuring deprivation, one has to necessarily accept some criterion on the basis of which social groups are distinguished in two or more categories. Naturally enough, the level at which categorization is made, is at the group level and not at the level of individual. The criterion (or, the classificatory variable), as such, entails the inevitable corollary that the behavioural variable also gets treated at the group level, and no valid inference can be drawn concerning the individual. *Prima facie* this seems to be all right, but it creates a situation which is likely to disturb the foundation of psychology itself.

The basic difference between psychology and sociology lies in the level at which behavioural data are analyzed. Sociologists analyze their data at the group level while psychologists are primarily concerned with the behaviour of individuals. If psychologists start using their antecedent and consequent variables at the level of groups alone without treating these variables at the individual level, the distinction between psychology and sociology is certain to disappear and psychology will subsequently lose its identity in the context of social psychological research. Thus, use of macroscopic approach denies the scope for predicting the behaviour of the individual, or explaining and controlling it. It is within this framework that measurement of the nature and extent of deprivation at the individual level becomes essential.

One must not be misled to the conclusion that the studies that have employed macroscopic level of measurement in investigations of deprivational effects have no psychological value. As a matter of fact, these studies are pioneers in the growth of psychological research in India. In any new area of research the initial sets of investigators are bound to be methodologically coarse. Methodological sophistication emerges as a function of insight that develops through sustained research in a particular area. It is only at this stage that psychologists are in a position to talk of microscopic treatment of deprivation. It is in this sense that one has to proceed from coarse-grain-analysis to fine-grain-analysis as, Sinha (1977b) advocates.

The Question of Optimum Experience

The most common strategies adopted in studying the role of life-situation-created deprivation-variables in the origin and growth of psychological functions are two: (i) to separate the deprived group from the non-deprived group; administer tests to them or record their performance under controlled conditions and statistically compare their average performance, or (ii) to determine the degree of deprivation for each subject as well as their individual performance and calculate correlation values between deprivation scores and scores on the chosen psychological measures. On the basis of obtained statistical differences or correlations it is usually concluded that deprivation leads to impaired performance or exerts significant influence on behaviour. It is in consonance with the most commonly assumed theoretical position that certain types of experiences or facilities are essential for proper growth of psychological functions. This assumption leads to a number of theoretical and methodological problems.

Let us assume that we have chosen some dimension or specific area, such as parental or cultural or nutritional deprivation along which an individual's deprivation level has to be determined. Let us further assume that we have arbitrarily decided to contain the entire dimension between zero and maximum degree of deprivation and divided it into seven equal appearing intervals envisaging increasing degree of deprivation. Now the most natural methodological question is what type of experiences of living facilities should be considered as involving zero degree of deprivation? This is in a way the problem of fixing the nature and kind of optimal experience. From purely methodological point of view its determination presents a knotty problem. Can it be determined in an absolute manner or has it to be a point relative to the conditions prevailing in those particular natural settings in which deprivation has to be measured? In either case, even though it may appear like hair-splitting, the determination of the zero point of deprivation or the level of optimal experience, seems to be a Gordian knot at the present stage of methodological know-how. This methodological problem is important enough to merit further elaboration.

It can be easily imagined that deprivation in any area of life is a bipolar dimension. At its one end lie those experiences which have to be assumed as maximally deprivational, while, at the other end, lie those experiences which are to be considered as maximally enriching. If it is so, it becomes a deprivation-enrichment dimension. Naturally, there must be a transition point where deprivation as well as enrichment are at a zero level. It is below that point that deprivation becomes more in amount and above which enrichment begins and gradually increases. The determination of this point is problematic as has been the case in measurement of attitudes. Since the problem has been tentatively resolved in the measurement of attitudes, it can also be done in measurement of experiences along the deprivation-enrichment continuum, provided that we adopt an approach which assumes that deprivation is an absolute phenomenon. However, the whole idea of deprivation-enrichment dimension becomes disrupted if one accepts deprivation as relative (Crosby, 1976). Within this theoretical framework experiential enrichment becomes a meaningless concept, because relative deprivation being more or less a comparative phenomenon becomes some kind of motivational variable and absence of such factors in any human being will amount to having zero motivational arousal level which can occur in sleep, coma and death only.

Measurement of Various Types of Deprivation

In the preceding discussion, we have hinted at terminological confusion prevailing in the realm of deprivation reflected in such terms as, of cultural deprivation, social and economic disadvantage, emotional, motivational, nutritional or parental deprivation. It can be safely assumed that these various types of deprivation imply differences in areas of deprivation involving lack of specified or specifiable kinds of stimulation patterns. Once we accept this position and by extending the generalization on the basis of results obtained from experimental investigations of sensory deprivation, to the areas of deprivation in natural setting, we hypothesize that the different types of stimulation patterns cumulatively received in specific areas, mentioned above, affect different types of psychological functions. As such, it becomes imperative that the experiential regions covered by these various types of deprivations be carefully mapped out and the boundaries of these regions be clearly demarcated. It is only then that measurement of various types of deprivation and their functional relations with specific psychological characteristics can acquire scientific respectability, and help in more minute and deeper analyses of such functional relationships.

Methodological Problems in Conducting Deprivation Studies

A review of psychological studies concerned with the effects of deprivation on cognitive, motivational and personality functions reveals that the current research strategies are fraught with a variety of methodological crudity and limitations which, more often than not, create contaminating and confounding effects. Therefore, the obtained results become more of artifacts than manifestations of genuine relationship. In this connection, two sets of problems can be isolated which, so far, have not received due attention. The first set of problems arise from the type of equipment, test materials and experimental tasks that are used in measuring the various functions and behaviours. The second set of problems arise from the kinds of investigators who conduct such studies and also from the kinds of respondents that are identified as deprived, especially under rural settings.

Equipments and Test Materials

Equipments and materials that we generally use for tapping perceptual, learning, problem-solving, representational or intellective functions are usually those that have been designed and developed by teachers and research scholars of psychology. The tests and equipments contain components which are mainly available to educated and richer segments of society that mostly live in fairly affluent conditions have acquired respectable degree of formal education. Since a majority of the deprived respondents come from rural background and are not well educated, it is self-evident that for them these test components and equipments are novel and distracting. Let us assume that we have to measure intelligence or capacity for depth perception or concept formation or any kind of cognitive skill in deprived persons. In any of these cases, we may use some standardized intelligence test, paper-pencil material containing pictures or outlines, memory drum, tachistoscope or some performance test, such as, Passalong test. In this assumed situation, we may follow one of the two strategies, *i.e.* either we bring the respondent to some psychological laboratory or we take the laboratory to the natural setting where our respondent lives. In the former case, not only the test equipment and materials are completely novel to the subject but the laboratory surroundings also are unfamiliar. It is a well known fact (Hebb, 1958; Bindra & Spinner, 1958; Berlyne, 1951; Dember, 1960) that novelty of material and situation can lead to performance decrement. Any difference in performance between two groups may not necessarily be due to deprivational factors but due to unfamiliarity of the respondents with the test equipments and materials. These methodological problems were realized long back. In 1964, Deutsch and his associates pointed out that standardized tests currently in use present three practical difficulties when they are used with disadvantaged minority groups: (i) they may not provide differentiation and the range of minority group scores, (ii) their predictive validity for minority groups may be quite different from that for the validation and standardization groups, and (iii) the validity of interpretations strongly depends upon an adequate understanding of social and cultural background of the group in question. Having realized these and other limitations, Sinha (1977 b) has made quite a few innovations and adaptations in preparation of test materials for use on deprived groups in Indian setting and has also stressed the need for culturally appropriate tools for data collection (Sinha, 1960, 1977a).

Investigator Elicited Problems

While doing research in one of the most backward areas of the country, namely the rural areas of eastern Uttar Pradesh, Misra and myself (Misra & Tripathi, 1980 ; Tripathi & Misra, 1975, 1976) were confronted with a variety of problems which threatened to thwart our research efforts in the very beginning. These problems were not specific to our research only. They are of generic importance and relate to the total set of interaction between investigators and respondents. Research investigators are usually Master's or Ph.D. degree holders in psychology or other social sciences, who are completely different in dress, diction and deportment from the respondents. They create the impression of being government officials or other types of organizational functionaries. This difference gives rise to very strong barriers in easy and smooth communication between the investigators and respondents. Because of this difference, the illiterate and experientially deprived members of the chosen community become suspicious, inhibited and evasive not only in answers to various questions put to them as part of interview schedule or test items, but also in making themselves available for various tests and experimental sessions. Even if the investigator is able to quell these initial hurdles by skilfully handling the problem of rapport, the initial difference in the three Ds mentioned above creates a feeling of respect for the investigator which vitiates or confounds the responses to test materials. This kind of confounding effect has been referred to as courtesy-effect by Doob (1968). This effect implies that such respondents are likely to give the interviewer the kind of response they believe he seeks rather than what they really believe in. Apart from these difficulties the well known response styles, such as acquiescence and social desirability, have been shown to vitiate such research situations to a large extent.

In the context of interaction between the investigator and respondent there is another factor which has not received as much methodological caution and care as it deserves. It is a well known fact that individuals coming from the deprived segments of society are deficient in vocabulary and verbal expression (Whiteman & Deutsch, 1968) and also in verbal learning with respect to hierarchy of verbal association, associative clustering, and syntactical mediation (Jensen, 1968). Hess and Shipman (1965) observed that 'the meaning of deprivation is the deprivation of meaning in a cognitive environment'. A number of authors such as Baratz and Baratz (1970), Goodman (1969), Labov (1970), Shuy (1969) and Stewart (1968), on the basis of their researches on poor children, conclude that such children come to school with a dialect which differs from standard English. This finding, which appears logically sound as wall as appealing to common sense, can be generalized more fruitfully to rural stratum of Indian society. If investigators are not well versed in the dialect of the respondents, quite naturally there would be communication gaps and misinterpretations. Under such circumstances, investigators are liable to get many more expected responses and thus become prophets of self-fulfilling prophecies, as is the case in a number of situations pointed out by Rosenthal and Jacobson (1968). Perhaps, on the basis of such confounding factors, one can explain the fact that most of the researchers working with human subjects drawn from deprived groups always obtain data that are expected on the basis of their hypotheses. Special care is required to have investigators who are efficient in the dialect of the respondents.

The Problem of Control in Deprivation Studies

As has been pointed out earlier, deprivation studies in natural settings are totally or partly correlational. The deprivation variable, no matter in what sense it is taken, is manipulated through some selection procedure and dichotomous or trichotomous groups of subjects are formed. These groups are subjected to experimental treatments or administered different psychological tests. When we select deprived subjects on the basis of this or that criterion, usually other factors also vary, and, consequently, interpretation of results on the basis of criterial variable becomes misleading. The problems of control have been ignored in those studies in which deprivation or social disadvantage has been treated as a categorical variable on the basis of caste (Rath & Dash, 1972) or caste and parental income (Das, *et al.*, 1970) or type of school and socio-economic status (Verma & Sinha, 1977) or age, caste and type of schooling (Sinha, 1977) or institutionalization (Sinha & Shukla, 1974) or residential area (Scares & Scares, 1969). In such studies, deprivation or disadvantage is assumed to be a direct function of some social or economic stratification. It does not require much effort to infer that even though deprivation may have positive correlation with the above mentioned criteria and many more, it does not necessarily follow that everyone belonging to a particular group or stratum is deprived. For example, let us take the case of parental income as the basis of separating or distinguishing the deprived from non-deprived. Imagine the case of a family which consists of three members consisting of father, mother and child with income of Rs. 200 per month and another family consisting of grandparents, parents and four children with an income of Rs. 500 per month. Obviously the former child would be considered economically deprived while the latter one becomes advantaged or non-deprived. In this case, naturally the family size is the crucial variable which has been left out uncontrolled. Experiences gained in a larger family have to be different from experiences gained in a smaller family, both in quality, as well as, quantity. Such examples are innumerable. The point worth bearing in mind is that such strategies in selecting samples of deprived and non-deprived subjects leave out many relevant variables that ought to be controlled or made a part of the strategy of deprivation assessment.

Even in those studies where deprivation is measured in the form of indices on the basis of a variety of criteria combined together (Rath & Samant, 1975; Misra & Tripathi, 1977a; Whiteman & Deutsch, 1968), the problem of controlling the relevant variables becomes very complex and involved, not only because studies in natural settings are essentially correlational, but because there is every possibility that different areas of life, where deprivation takes place, interact among themselves. In the first instance, since these researches are conducted using the R-R paradigm and deprivation is the independent variable *i.e.* the first R in the paradigm, and other innumerable R variables become relevant to the variations in the second R. The first R variable is manipulated through selection, other R variables may be effectively controlled, but this is not always the case. In many studies where cognitive processes are studied as functions of deprivation level, personality variables which have been undoubtedly shown to affect cognitive functions are seldom controlled, though, personality too has been shown to be a function of deprivation. On the other hand, it can also be asserted that deprivation-areas interact with one another even though it has not yet become clear as to what is the exact nature of this interaction. If it is true, it becomes self-evident that picking up the deprived and non-deprived groups on the basis of one life area alone, such as, economic or maternal deprivation, does not provide scores for matching the two groups of subjects in terms of the experiential content in relation to other areas of deprivation. And, therefore, the probability remains very high for sources of error to become systematic without the investigator's knowledge. For precision, accuracy and validity, it is necessary that we direct our efforts, towards more microscopic treatments and analyses in deprivation studies.

References

Adams, A.A. (1971). The effects of planned background experiences on economically disadvantaged kindergarten children. Unpublished Doctoral Dissertation, Missisippi State University, Missisippi.

Baratz, S.B., & Baratz, J.C. (1970). Early childhood intervention: The social science base of institutional racism. *Harvard Educational Review, 40,* 29-50.

Bee, H. (1974). *Social Issues in Developmental Psychology*. New York: Harper and Row.

Beriyne, D.E. (1951). Novelty and curiosity as determinants of exploratory behaviour. *British Journal of Psychology, 41,* 68-80.

Bindra, D., & Stpinner, N. (1958). Response to different degrees of novelty: The incidence of various activities. *Journal of Experimental Analysis of Behaviour, 1,* 341-350.

Bladslcc, A.L. (1967). Nutritional time bomb. Today's Health, *45,* 7. Coleman, J.C. *Psychology and Effective Behavior*. Bombay: D.B. Taraporevala.

Crosby, F.A. (1976). Model of egoistical relative deprivation. *Psychological Review, 83*, 85-113.

Das, J.P. (1973). Cultural deprivation and cognitive competence. In N.R. Ellis (Ed.) *International Review of Research in Mental Retardation* (Vol. 6), New York: Academic Press.

Das, J.P., Jacbuk, K., & Panda, T.P. (1970). Caste, cultural deprivation and cognitive growth. In H.C. Haywood (Ed.) *Social Cultural Aspects of Mental Retardation*. New York: Appleton-Century-Crofb.

Das, J.P., & Pivato, E. (1977). Malnutrition and cognitive functioning. In N.R. Ellis (Ed.) *International review of research of mental retardation*, (Vol. 8), New York: Academic Press.

Davis, R.E. (1968). Social deprivation and cognitive functioning in lower class children. *Dissertation Abstracts, 29*, 1105-1406.

Deutsch, M., Fishman, T.A., Kogan, L.S., North. R.D., & Whiteman, M. (1964). Guidelines for testing minority group children. *Journal of Social Issues, 20*, 129-145.

Deutsch, C.P. *(1965). Effects of environmental deprivation on basic psychological processes: Some hypotheses*. Paper presented at the annual meeting of American Educational Research Association, Chicago, Illinois.

DeBoard, L.W. (1968). The achievement syndrome among Negro and White culturally disadvantaged boys. Unpublished Doctoral Dissertation, Vanderbilt University.

Dember, W.N. (1960). *The Psychology of Perception*. New York: Holt.

Dobbing, J. (1967). Growth of the brain. *Science Journal, 5*, 81-86.

Doob, L.W. (1968). Just a few of the presumptions and perplexities confronting social psychological research in developing countries. *Journal of Social Issues, 24*, 71-81.

Getzels, J.W. (1969). The social psychology of education. In G. Lindzey, & E. Aronson, (Ed.) *The Handbook of Social Psychology*, (Vol. 5), New York: Addision Wesley.

Gewirtz, J.L. (1961). Learning analysis of the effects of affective privation in childhood. *Acta Psychologia, 19*, 404-405.

Goodman, K.S. (1969). Dialect barriers to reading comprehension. In J.C. Baratz & R.W. Shuy (Eds.) *Teaching Black Children to Read*. Washington D.C.: Centre for Applied Linguistics.

Goff, R.M. (1954). Educational implication of the influence of rejection on aspiration levels of minority group children. *Journal of Experimental Education, 23*, 179-183.

Goldfarb, W. (1945). Psychological privation in infancy and subsequent adjustment. *American Journal of Ortho-psychiatry, 15*, 247-255.

Gordon, J.E. (1968). The disadvantaged pupil. *Irish Journal of Education, 2*, 68-105.

Green, L., & William, W.E. (1965). Negro academic motivation and scholastic achievement. *Journal of Educational Psychology, 56*, 241-243.

Green, R.L., Hofmann, L.J., & Morgan, R. (1967). Some effects of deprivation on intelligence, achievement and cognitive growth. *Journal of Negro Education, 36*, 5-44.

Havighurst, R.J. (1964). Who are the socially disadvantaged ? *Journal of Negro Education, 33*, 210-217.

Hebb, D.O. (1958). *A Text Book of Psychology*. Philadelphia: W.B. Saunders.

Hess, R., & Shipman, V. (1965). Early experiences and the socialization of cognitive modes in children. *Child Development, 36*, 869-885.

Jensen, A.R. (1968). Social class and verbal learning. In M. Deutsch, I. Katz & A.R. Jensen (Eds.) *Social Class, Race and Psychological Development*, New York: Holt, Rinehart & Winston.

Kogan, J., & Kogan, N. (1970). Individual variation in cognitive processes. In P.H. Mussen (Ed.) *Carmichael's Manual of Child Psychology*, (Vol. 1), New York: John Wiley.

Labov, W. (1970). *The Study of Nonstandard English*. Illinois National Council of Teachers of English.

Langmeier, J., & Maiezeck, Z. (1968). *Psychological deprivation in childhood*. Praha: SZN.

Langmeier, J. (1972). Personalities of deprived childrer. In F.J. Monks (Ed.) *Determinants of Behavioural Development*. New York: Academic Press.

Langmeier. J., & Matezeck, Z. (1975). *Psychological Deprivation in Children*. New York: Hoisted.

Miller, J.O. (1968). *Cultural Deprivation and its Modification: Effects of Intervention*. Paper presented at the International conference on 'Socio-cultural aspects of mental retardation'. Nashville.

Misra, G., & Tripathi, L.B. (1977a). The concept of prolonged deprivation and its measurement. *Indian Journal of Behaviour, 1*, 48-60.

Misra, G., & Tripathi, L.B. (1977b). *Manual of Prolonged Deprivation Scale*. Agra: National Psychological Corporation.

Misra, G., &Tripathi, L.B. (1980). *Psychological Consequences of Prolonged Deprivation*. Agra: National Psychological Corporation.

Misra, G., & Tripathi, L.B. (1978). Prolonged deprivation scale. *Indian Journal of Psychology, 53*, 65-76.

Munsey, W.L. (1971). Development of an index of cultural deprivation independent of ethnic group membership. *Dissertation Abstracts, 32*, 3051.

Nurcombe, B. (1970). Deprivation: An essay in definition with special consideration of the Australian aboriginals. *Medical Journal of Austria, 2*, 88-92.

Rankin, N.O. (1970). The effects of selective social deprivation in & natural setting. *Dissertation Abstracts, 30*, 4820-4821.

Rath, R., & Dash, A.S. (1972). *Cognitive Growth and Class Room Learning of*

Naturally Deprived Children in the Primary Schools. Paper presented at the East West Center for Cross-Cultural Studies, Honolulu.

Rath, R., & Samant, C. (1975). Cultural deprivation index. Unpublished manuscript Department of Psychology, Utkal University, Bhubaneshwar.

Riessman, F. (1962). *The Culturally Deprived Child*. New York: Harper & Row.

Rosenthal, R., & Jacobson, L. (1968). Self-fulfilling prophecies in the classroom: Teacher's expectations as unintended determinants of pupil's intellectual competence. In M. Deutsch, I. Katz & A.R. Jensen (Eds.) *Social Class, Race and Psychological Development*. New York: Holt, Rinehart & Winston.

Rossi, A.M. (1969). General methodological considerations. In J.P. Zubek, (Ed.) *Sensory Deprivation: Fifteen Years of Research*. New York: Appleton-Century-Crofts.

Shuy, R.W. (1969). Linguistic background for developing reading materials for black children. In J.C. Baratz, & R.W. Shuy (Eds.) *Teaching Black Children to Read*. Washington D.C. Centre for Applied Linguistics.

Singh, A.K. (1976). *Social Disadvantage, Intelligence and Academic Achievement*. Paper presented at the 14th annual conference of the Indian Academy of Applied Psychology, Indian Institute of Technology, Kharagpur.

Sinha, D. (1969). *Indian Villages in Transition: A Motivational Analysis*. New Delhi: Associated Publishing House.

Sinha, D., & Shukla, P. (1974). Deprivation and development of skills of pictorial depth perception. *Journal of Cross-Cultural Psychology, 5*, 434-450.

Sinha, D. (1977a). Social disadvantages and development of certain perceptual skills, *Indian Journal of Psychology, 52*, 115-132.

Sinha, D. (1977b). Orientation and attitude of the social psychologist in a developing country: The Indian case. *International Review of Applied Psychology, 26*, 1-10.

Scares, A.T., & Scares, L.M. (1969). Self perceptions of culturally disadvantaged children. *American Educational Research Journal, 6*, 31-45.

Stewart, W.A. (1968). Continuity and change in American Negro dialects. *The Florida Reporter*, Spring.

Stot, L.H. (1974). *The Psychology of Human Development*. New York: Holt, Rinehart.

Suchman, E.A. (1966). Medical deprivation. *American Journal of Orthopsychiatry, 36*, 665-672.

Symmonds, M. (1968). Disadvantaged children growing in a climate of hopelessness and despair. *American Journal of Psychoanalysis, 28*, 15-24.

Tripathi, L.B., & Misra, G. (1975). Cognitive activities as a function of prolonged deprivation. *Psychological Studies, 21,* 54-61.

Tripathi, L.B., & Misra, G. (1976). Some cognitive processes as functions of prolonged deprivation. *Indian Journal of Psychology, 51,* 129-143.

Talmadge, M., Hayden, B.S., & Schiflf, D. (1969). Longitudinal analysis of intellectual and educational achievement change in culturally deprived and emotionally disturbed boys. *Perceptual and Motor Skills, 29,* 435-440.

Tulkin, S.R. (1972). An analysis of the concept of cultural deprivation. *Development Psychology, 6,* 326-339.

Uzgiris, I.C. (1968). *Socio-cultural Factors in Cognitive Development*. Paper presented at the International conference on 'Socio-cultural aspects of mental retardation'. Nashville, Ten A.

Verma, L., & Sinha, M.M. (1977). Effects of SES on perceptual differentiation contingent upon induced frustration in children. *Indian Psychological Review, 15,* 27-35.

Whiteman, M., & Deutsch, M. (1968). Social disadvantage as related to intellective and language development. In M. Deutsch, I. Katz, A.R. Jensen (Eds.) *Social Class, Race and Psychological Development*. New York: Holt, Rinehart & Winston.

Wight, D.W. Cloniger, M.F., & Kneene, P.J. (1970). Cultural deprivation: Operational definition in terms of language development. *American Journal of Orthopsychiatry, 40,* 77-86.

Wilcox, R. (1971). *Psychological Consequences of Being a Black American*. New York: John Wiley & Sons.

Yarrow, L.J. (1961). Maternal deprivation: Toward an empirical and conceptual re-evaluation. *Psychological Bulletin, 58,* 459-490.

7

Some Erroneous Beliefs of Parents about their Children

It is universally known that every human being is a product of his heredity and environment, and parents are the sources of both especially during the critical periods of growth and development. Heredity provides developmental potentials and parental actions and reactions provide opportunities as well as resources for actualization and development of genetic potentials. Proper development or developmental aberrations are primarily determined by parental expertize, skills, attitudes and behaviours. Proper and efficient upbringing of a child is a highly complex task. However, it is paradoxical that in India at the macro-level this complexity is neither fully realized nor even appreciated by most of the educated peoples. The paradox is that for the job of primary or nursery school teachers formal training, is considered mandatory, but for becoming efficient parents no training in parenting is required nor provided for. Parents in this country bring-up their children on the basis of their own childhood experiences, folklore wisdom and close observation of other parents etc.

Parents, unaware of the psychology of child development and not having learnt the skills for effective parenting, develop some erroneous beliefs about the nature and characteristics of the psychological needs of children. Most of these beliefs and parental behaviours guided by them, have deleterious effects on child development. These beliefs are numerous and most of the parents share them. It would be useful to mention some of them.

It is commonly believed that children must obey their parents and must not ever question or disagree with their parents and superiors. Such a belief interferes with the development of self-esteem, self-confidence and sense of independence. Parents do not always maintain a clear-cut distinction between a child and its behaviour, because they implicitly believe that a child and his behaviours are the same. They do not realize that non-desirable or non-comforming behaviours may be due to ignorance or lack of skill; emotional disturbance or non-fulfilment of some deficiency needs. Bad behaviours do not mean that the child is bad. It is the behaviour which is inappropriate and violative of social norms. One of the most harmful and subtly erroneous beliefs is that children can upset or hurt the feelings of parents. Here even highly educated parents do not make a discrimination between physical and psychological hurts. A child's actions and reactions are not the direct agents of hurt. It is the perception of action by the father or mother which becomes hurtful. It simply means that the parents have not yet grown up enough to perceive child's behaviour in a correct perspective.

Barring a few exceptions all parents punish their children for indulging in undesirable and grossly norm violating behaviours. They do so under the assumption that punishment, blame and guilt are effective methods of managing children. Instead of correcting children this method has more adverse effects. Punishment does not correct under any condition. It either temporarily suppresses behaviour or strengthens those behaviours in case of mild punishment. However, punished children gradually develop negative attitudes and quite often hostility towards the punishing agents. Frequent and heavy punishments create inerasable scars on a child's psyche. Some parents on the other hand, erroneously subscribe to the idea that praise spoils children and under normal conditions of interaction they must not be praised. If the child does well, some parents take it for granted and think that the child has done what is expected of him/her. Such an attitude makes children cold, isolated and somewhat apaihetic.

Some parents are over protective and quite often pamper them because of their tacit belief that children must not be frustrated. They do not realize that occasional frustrations are not only natural facts of life, but also motivation enhancing. If properly handled, frustrated children learn to work successfully in the face of challenges. Another common belief, prevalent among some well educated parents is that if a child misbehaves, it must be punished heavily and immediately. It is true that severe punishment suppresses behaviour for longer period but creates highly undesirable effects in a child. It creates not only emotional problems but lays foundations of cheating, telling lie and becoming negativistic..

All parents are quite frequently frustrated by their children. On such occasions parents as well the child are upset. In such cases some parents believe that the child must be calmed first and then they will naturally be calmed. *Prima facie,* such an attitude appears to be logical, but really it is an inaccurate way of looking at it. When children fail in some test and feel frustrated, many parents get upset which is contagious and upsets the child. In this situation the best thing for the parents is not to get upset because their child has failed and feels frustrated. As a matter of fact each failure should be treated as natural and should be harnessed as an extra motivational force.

The most damaging belief, parents have but are not fully aware of, is that children learn more from what their parents, teachers and/or guardians tell them to do rather from what they themselves do. It is true that in early childhood they learn a lot by trial and error and contiguity followed by success, reward and praise by elders. But such learning is extremely limited in scope. From childhood and onward children learn by modelling or observation or imitation. As a matter of fact no one learns anything by preaching. Let us take an example. Innumerable parents complain that their children do not study in the evenings, do not get up early in the morning and study, and waste their time in extraneous activities. Such parents do not look at themselves and see as to how do they spend their times. They should ask: do they practice what they preach ? Children develop habits and reaction pattern on the basis of what they see their parents and superiors doing.

Untrained and naive parents mishandle their children and do not understand proper patterns of actions and reactions towards them. More often than not father's reactions to a child's misconduct are incongruent. Suppose a father is scolding his son for some misconduct, but the mother intervenes and protests against his scolding the son. Such incongruent behaviours create confusion in children and they fail to learn discrimination between what is right and what is wrong. Lack of synchronization, manifested by parents in their actions and reactions towards their offsprings, has numerous deleterious effects on child-development giving rise to a variety of behavioural aberrations. Father's pattern of kindness-unkindness and firmness-nonfirmness may be opposed to that of the mother culminating into maladaptive development of a child. The best pattern is to be kind and firm, and this pattern has to be followed by both the parents.

Behaviours of parents emanating from the aforesaid erroneous beliefs and incongruent parings of kindness-unkindness and firmness-nonfirmness give rise to maladjustive development in the offsprings, such as fear of the darkness because an abnormally treated child imagines existence of evil or harmful entities lurking in darkness; fear of strangers, which culminates into extreme shyness, fear of failure which undermines emergence and growth of achievement motivation and kills the potential for taking initiatives when confronted with problems, fears of injury, rejection and ridicule are common symptoms found in children who are brought up in a lopsided manner.

One of the most important purposes of bringing up children is to socialize them in such a way that they learn to control their emotions and their raw expressions, especially highly negative emotions such as anger or temper-tantrum anxiety and depression. Emotional expressions are innate and occur naturally, while thinking and reasoning develop gradually as a function of maturation as well as schooling, irrespective its being formal or informal. In spite of this, children, if properly brought up, gradually learn to control impulsiveness and anger. The dynamics of anger are rather surprisingly simple. A child becomes angry or enraged if it does not get what it desires. It is so because non-fulfilment of desire leads to frustration that gets transformed into anger, bitterness as well as aggression. Under such conditions parents have to play a difficult role. In order to control child's anger one has to maintain one's cool and guide the childs how it can save itself from the travails of frustrations and learn to control reactions in a rational manner.

As the children grow, areas of their interaction with children of neighbourhood and of school becomes expanded. As a consequence of interactional expansion children learn incidently to misbehave and develop undesirable habits. Few parents understand the reasons that underly such behaviours. All growing children want attention, power, revenge or to be left alone by showing some forms of disability. Effective skill parenting requires that both father and mother along with the other older member in an extended family to address this question to oneself: What the child is trying to achieve by his/her misbehaviours? What is the goal? Parents seldom ask such questions.

The least spectacular problem but most challenging task facing parents is teaching self-discipline or self-regulation. How to inculcate the spirit of self-discipline, a sense of responsibility in one's child, calls for highly complex behavioural skill.

So far an attempt has been made to emphasize a simple point. How parents behave in general and how they act and react to their children are the most important determinant of a child's psychological development. For proper effective development of children it is essential that parents must be well-versed in the principles of developmental psychology, be good models and must have the necessary skills of acting and reacting to ward their children. Unfortunately parenting has so far in this country has not been recognized as a technically sophisticated job. It is therefore necessary to create awareness in the masses about the need of effectively skilled parenting or not only making one's family life more peaceful, comfortable and happiness enhancing, but also for creating better society in the country for the future. Once this awareness is created, there shall be widespread demand for training programmes in the scientific techniques of effective parenting.

8

A Theoretical Perspective on Man-Environment Relationship

Today it is intensely realized that the prevailing man-environment relationship is extremely disturbed. Man is proving to be detrimental to environment and in return environmental disturbances are causing devastating effects on human development. It is needless to point out that the current man-environment relationship is rooted in erroneous conceptualizations of man as well as of environment owing to which organized human activities and socio-political interventions in environment have raised the level of environmental pollution, disturbed the ecological balance and have led to rapid corrosion of environmental structure to an alarming state.

In this backdrop it is crystal clear that there is an urgent need for reorientation and re-thinking regarding the conceptual framework of, and meta-theoretic assumptions regarding the nature of environment as well as of man. It is a well-known fact that the contemporary environmental psychology subsists on concepts and constructs borrowed from biological ecology, ethology, social pathology and anthropology. When used in behavioural contexts, these concepts and constructs exhibit lack of logical inter-relatedness and operational specificity. They are also based on untenable meta-theoretical assumptions. Psychologists in general and environmental psychologists in particular, in their zeal to fashion the discipline of psychology in accordance with the models of physical and biological sciences, have turned blind eyes to some extremely significant features of human functioning and environmental properties. Such a state of affairs in psychology has not only rendered it an ineffective branch of knowledge but has also hindered in developing an ideal model of man-environment interaction.

The Construct of Environment

The construct of environment is indeed ubiquitous. It is used both in biological and social sciences. Each discipline defines it according to its requirements and theoretical perspectives. Irrespective of its use in various disciplines quite frequently, it is employed as a highly general term and includes the concepts of eco-system or biomas, ethos, habitat, social settings and prevailing social, political and/or economic conditions. In biological ecology and community psychology environment is conceptualized as having four easily distinguishable features. First, it is asserted that all components of environment are intrinsically *inter-dependent*. It means that if and when one component is changed or activated, alterations between all other components occur automatically (Kelley, 1971; Mills & Kelley, 1972). Its second feature is *cycling of resources* implying thereby that energy discharged by one component is received and transformed by other components in a cyclic manner. On the other hand, all components, especially living organisms, possess inherent features of wide-range of *adaptability* or capacity for tolerance of fluctuations in environmental functioning. Only drastic changes in environmental structures and functions owing to natural disasters or man-made ruinous interventions in nature prove injurious to various components and jeopardize the existence of less adaptable components. The most conspicuous feature of environmental functioning is its guidance by *the principle of succession*. It refers to the deterministic nature of environment. Whatever changes occur in it are directional and predictable. Modification of the physical environment culminates in the establishment of as stable an eco-system as is biologically possible on the site. Because of these four features of environment it has been possible for biological and physical sciences to succeed in explaining, controlling and predicting environmental phenomena.

Environment as a construct in psychology has been extensively employed as one of the two major determinants of psychological processes and behaviours. Notwithstanding, the fact that there is continuous interaction between man and environment, and each is conceived as affecting the other, the construct of environment has not been analyzed and characterized with as much seriousness as it should have been. Whether it is extreme environmentalism of Watson, or the concept of 'press' in Murry's personalogy or the behaviour considered as a joint function of person and environment in Lewin's dynamic psychology, interactional features of environment have not been incorporated adequately in its conceptualization. With the origin and growth of environmental psychology recently, conceptualization and measurement of environmental differences have come to the fore. However, even today psychologist's emphasis on the variance of behaviour as a function of interaction between personal and environmental variables cannot be considered enough for understanding either human behaviour or environmental features. In order to understand human behaviour and psychological consequences of man-environment interaction, it is imperative to conceptualize environment as an entity determining human behaviour as well as being continuously modified by human behaviour. It is not enough to interpret environment as: (1) stimulus source, (2) behavioural context, and (3) resource of need fulfilment only, but also as a dynamically changing entity.

Keeping in view the shortcomings of the contemporary conceptualization of environment in behavioural sciences, alarmingly increased environmental pollution, threateningly disturbed ecological balance, appearance of holes in the ozone, shifts in the climate all over the world and unavoidable interventions in environmental structures for development of human societies, one has to take a fresh look at the environment. Now it can be postulated that *environment in incessant interaction with human beings is a multi-faceted and complex entity which is both distal as well as proximal and which possesses the properties of being active in the form of providing stimulations, behavioural contexts and incentives, and of being reactive to human actions and interventions, is in constant flux.*

Eventhough it is a somewhat cumbersome definition, it brings out a number of functional features of environment.

Environment is multifaceted in the sense that it is natural having mountains, lands, rivers, oceans, lakes, flora and fauna, weather, seasons, air currents, storms and rains etc. It is also man-made, such as roads and buildings, cities and villages, gardens and forests, canals and dams, railway lines and smoke belching mills and factories, noise producing loud speakers and booming planes etc. It is complex not only in the sense of having innumerable interactive or interrelated entities, but also in the sense that it varies along the concrete-abstract dimensions at physical, biological, political, economic, and psycho-social levels. Environment is distal because it is there existing objectively as something real. It is proximal in the sense that human beings consider them as perceived by them. It is the proximal or the perceived environment to which human beings act or react.

Human actions and reactions, whether made on large scales by organized groups or on small scales by individuals, occur as interventions and affect the environmental components and their functions. For example, suppose in some locality there is dearth of water for irrigation purposes. As a reaction to this water scarcity human beings set up tube wells or build canals through which river water is channelised to farm and fields. Naturally, as a consequence of such efforts there is increase in moisture and humidity, and waterlogging in low lying areas culminating in enhanced mosquito menace. Large scale interventions quite frequently disturb ecological balance in the areas and create new problems. Needless to point out that deforestation and though less felling down of trees for agricultural and other purposes, killing of wild animals, dumping of industrial waste products and flow of sewers in rivers have created problems of environmental pollution, and disturbed ecological balance. Environment does not tolerate interventions by remaining passive and inert. It becomes automatically active to absorb the changes in its structure created by interventions and transforms itself to re-establish its balance. Such environmental transformations occur in a deterministic manner and can be predicted.

Environment unlike human beings possess no intentions and cannot be considered as either malevolent or benevolent but as valuefree. It is the human thinking that perceives and evaluates environmental features as beneficial or injurious in accordance with human needs and attitudinal predispositions.

Thus, environment is constantly in a flux of deterministic cycles of activities. Whenever interventions occur in its functions through modification of its structural components, it becomes extremely reactive and modifies its other components to re-establish deterministic nature of continuous flux of cyclic activities.

The Nature of Human Beings

In man-environment interaction the man as an actor has decidedly not yet been properly conceptualized. It is true that in the science of psychology there has been numerous efforts to delineate and delimit the inherent nature of human beings and a variety of meta-theoretic assumptions have been made. Under the pervasive influence of Darwinian biology and the logical positivism behaviourists accepted infrahuman species as simple models of man, and adopted experimental observation as the method of investigation. They assumed that human beings were basically passive entities under the control of external stimulation and their behaviours were conceived as reactions or responses to environment. Thus human beings were considered as reactive and associative organisms to be studied in the framework of learning paradigm. Cognitivists, such as Tolman and others, generated powerful experimental data against the behaviouristic model of man. Under the influence of cybernetics and information processing approach cognitive psychologists have conceptualized human beings as active, organizing, activity-initiating and self-regulating. Accordingly, self-regulation is analyzed and interpreted in terms of self-generated behavioural standards, self-evaluation, comparison and actional consequences (Kanfer, 1977; Kanfer & Hageman, 1981).

Such meta-theoretic assumptions regarding the inherent nature of human beings are indeed quite effective for interpretation of a variety of psychological data, but they do not clarify the intricacies of man-environment interaction and psychological as well as environmental consequences of such interactions on a macro-level. There assumptions are based on (or limited by, if you would) scientism, individualism-universalism and causal forms of explanations (Harre, Clarke & Carlo, 1985). In late fifties a number of psychologists, such as, Carl Rogers and Maslow, found the meta-theoretic underpinnings of then prevalent science to grasp the inherent nature of human beings. Having become disillusioned with the main stream of psychological thinking they proposed a new type of psychology, usually called humanistic psychology and considered as the third force.

The humanistic approach to the study of behaviour focuses not only on what the person, is but also on what the person has potentials to become. This new emphasis on the inherent potentials for becoming enables psychology to conceptualize human nature beyond the narrow limits of scientism and at the same time provides theoretical grounds for explaining numerous human activities which were once considered beyond the domain of psychological explanations. Such an approach to the study of human beings underscores the distinction between science and ethno-science (Weiner, 1980).

The humanistic emphasis on potential for becoming not only takes into consideration the growth needs of human beings, but it also provides a conceptually sound basis of understanding and analyzing a variety of human activities which cannot be validly explained in terms of cognitive theories and other psychological approaches. Human beings do not interact with the environment just to fulfil their survival needs. They do so to attain mastery over some aspects of their environment and thereby enhance and enlarge their self-esteem. It is self-evident that survival needs orient human beings to their environments and make them dependent on their physical as well as socio-economic surroundings, while growth needs provide motivational support for manipulating environmental components and for attaining mastery over them. The construct of growth need is substantiated by a number of cultural activities which characterize all societies in this world.

At this juncture an important question must be raised. Why do human beings build mansions, create music and complicated dance forms, and build bridges and factories? In order to answer this question higher order needs have to be posited.

When culture is defined as a process through which human beings improve upon nature and natural resources in order to make life more comfortable, more enjoyable and aesthetically more enriched, it becomes clear that culture as a process is somewhat opposed to the processes of nature. Nature or components of the natural environment are guided by the principle of least effort or resistance; and natural events occur with maximum spontaneity in a deterministic manner. On the other hand, cultural processes call for best efforts and resistance. Human beings, because of demands of the cultural processes, have to follow the principle of best effort in every area, be it scientific or literary, artistic or technological, in order to achieve mastery, excellence or competence. Naturally enough, the motivational force behind cultural activities must be the need for mastery and self-esteem. It is because of this urge that man endeavours to manipulate environmental components.

The survival needs make human beings accommodative and adjustive to the constraints of the physical environment, but as they grow and attain optimum assurance of the fulfilment of survival needs, growth needs impel them towards cultural processes. They manifest behaviours that are guided by the principle of least effort, and instead of remaining passively adaptive, human beings become actively manipulative, interventive and quite often exploitative in accordance with their genetically created and environmentally nurtured capacities and capabilities for growth and self-enhancement.

Considered in the aforesaid perspective it can be postulated that *inherent nature of man is determined, controllable and predictable to the extent it is subject to environmental constraint and control,* but because of the naturally occurring developmental processes, *man is also inherently disposed to move in the direction of becoming free and unpredictable to the extent it becomes self-regulated, capable of successfully manipulating his environment and is able to rise above temporal spatial constraints in developing his self and self-control.* The entire process of life-span human development may be viewed as a directional movement from dependence to independence, from environmental adaptation to self-regulation and from determinism to free choice and voluntary self-control.

Man-Environment Interaction

Keeping in view the basic nature and features of environment and of human beings some pertinent questions merit serious discussion. What kind of man-environment interaction is there at present in this world? Is the prevailing interaction good enough to sustain each other? If not what is wrong with it? Is it possible to change it and to promote healthier relationship which may be considered mutually beneficial? If so, how can such a relationship be promoted?

From human point of view man-environment interaction can be categorized under two broad headings, *i.e. adjustive-accommodative* and *manipulative-exploitative*. It is self-evident that in adjustive interactions human beings treat environment as a fountain of those resources which are necessary for fulfilling survival needs and are easily available in the environment. Human beings adjust their needs to suit the environmental forces and use it to the extent that there is no design for harnessing and guiding the natural forces. Experienced regularities in environmental features and phenomena shape the human strategies for fulfilling survival needs. In this kind of relationship there is no human effort to subjugate nature but to live with it. Houses are built on sites that are beyond the reach of flood water. Crops are sown according to natural weather, climate and season. In such interactions there is a predominant element of dependence on nature with minimal amount of manipulation, and maximal amount of accommodation.

On the other hand, following manipulative-exploitative interaction dams are build to prevent damage by flood, predators and wild animals are killed to save lives of domesticated animals, to meet food and leather requirements, and to satisfy survival and growth needs. Forests are cleared to enlarge areas of cultivation and to meet timber requirements. Roads, factories and electricity producing centres are build. Coal, iron ore and other minerals are extracted through mining and metallurgy technology. Irrigation canals are made to make life more comfortable and to satisfy the need for mastery over the environment. In such interactions there is maximal human effort for subjugating nature and harnessing it to the welfare of the mankind.

It is self-evident that the prevailing interactions of man with environment are exploitative-manipulative. It is also evident that this type of relationship has led to extremely disastrous modifications of biomas and strife-ridden fast depletion of natural resources, dangerously polluted and disbalanced eco-systems. Contemporary social ecological awareness has reawakened people all over the world for having a fresh look at this kind of interaction. Environmental sciences have now become one of the most popular concerns of the governments, bureaucrats and intelligentsia. There is unprecedented upsurge of 'save the environment' campaign all over the world. It is now intensely realized in certain quarters that exploitative interaction without proper regenerative skills has caused unprecedented damage to environment. It is also felt that a distinction between manipulative and exploitative interactions is urgently needed. In manipulative relationship the manipulator has necessary concern for the continuous existence of the manipulated, while in exploitative interaction such concerns are altogether absent and the exploiter focuses on his gains with no concern for the kind of injury his exploitation inflicts on the exploited. The modern man, in his relentless pursuit of affluence, convenience, comfort and mastery over his surroundings, has fully exploited the natural resources and has used his environment as the dumping ground for enormous amount of industrial waste and garbage. The result is that the entire mankind has found itself on the verge of worst environmental crises.

Ideal Man-Environment Relationship

Now the most pertinent question that is being raised is: what should be the most desirable relationship between man and environment? There is no denying the fact that interaction between the two is a necessary condition of human existence. Keeping in view the inherent nature of human beings and the active-reactive properties of environment it has to be accepted that man has to manipulate the components of environment be it natural, economic or psycho-social; and it has also to be accepted that environment has to react to those manipulations in order to maintain its existential balance. Environment has to simultaneously provide stimulation, incentives and behavioural contexts. In this backdrop one can think of only one type of relationship that can be accepted as the most desirable. It has to be symbiotic relationship as opposed to antibiotic one. It implies a relationship in which human manipulations are such that they not only provide opportunities for fulfilling survival needs but also the growth or cultural needs, and at the same time sustain and enrich the existence of environment in such a way that the pristine purity of environment is not polluted, ecological balance is fully maintained and man-made environment does not annihilate the naturally occurring environmental phenomena and processes. In other words, human manipulation has to be such that the environment does not become over-reactive for maintaining its balanced existence.

Prima facie the idea of symbiotic relationship appears as an impossible proposition. How can one be manipulative and nurturant at the same time? However, a deeper analysis of symbiotic relationship makes it not only a practicable scheme but also a highly self-actualizing strategy. If one delves deep in the ancient pattern of Indian thinking one is certain to get some useful glimpses of man's symbiotic relationship with nature. Ancient Indians tilled lands for agricultural purposes, dug wells, built forts and temples, extracted minerals for making implements, tools, weapons as well as ornaments. They also hunted animals for food; killed predators and wild animals, and simultaneously endeavoured to co-exist with nature. For them most of them phenomenal components of the natural environment were regarded as abodes of deities. All rivers were considered as benevolent mother-like deities, most of the trees were considered as the seats of some supernatural beings. Every *peepal* tree was respected because a deity called *Vasudeva* lived there. *Neem* tree was regarded as the place of Goddess *Kali*. All trees symbolized unperishable continuity of life. Fire, water, air, earth and space were deities of *Agni*, *Varun*,

Pavan, *Vasundhara* and *Indra*. Man, thus, exhibited profound respect for nature. Exploitation of nature was considered as a sin and was strictly prohibited. To dig a well or to get a pond constructed, to plant trees and to keep the air clean by performing *yagya* were considered as self-enhancing religious duties (*dharma*) of man. Man existed in harmonious relationship with nature.

The ancient Indian ways of treating nature is cited here to illustrate the feasibility of symbiotic or mutually sustaining man-environment relationship. It is in no way suggested that the world should take a retrogressive step and accept the ancient Indian percepts. What is suggested is to change the human perception of environment and treat it not only as a source of stimulation, incentives and behavioural context for fulfilment of human needs of survival and growth but also treat it as an active and powerfully reactive agent. It has to be treated with the same kind of respect and restraint as one has to treat social groups in social interaction.

How can such an interaction be developed? It is a tricky question. In order to develop mechanisms and strategies for promoting symbiotic relationship between man and environment, global efforts are to be made on an international level involving all nations. Galloping consumerism has to be stopped, goals of technological development to be redefined, natural resources to be utilized with foresight in a restraint manner, and population growth has to be checked more effectively. Unless such an effort is immediately initiated, the very existence of mankind is likely to be totally annihilated by environmental reactions to contemporary exploitations of its resources.

References

Harre, Rom, Clarke, David & Carlo, Nicola De (1985). *Motives and Mechanisms: An Introduction to the Psychology of Action*. London and New York: Methuen.

Kanfer, F.H. (1971). The maintenance of behaviour by self-generated stimuli and reinforcement. In A. Jacobs & L.B. Sachs (Eds.) *The Psychology of Private Events*. New York: Academic Press.

Kanfer, F.H. (1977). The many faces of self-control, or behaviour modification changes its focus. In R.B. Stuard (Eds.) *Behavioural Self-Management: Strategies, Techniques, and Outcomes*. New York: Brunner/Mazel.

Kanfer, F.H. & Hageman, S. (1981). *The role of self-regulation*. In L.P. Rehm (Ed.) *Behaviour Therapy for Depression, Present Status and Future Directions*. New York: Academic Press.

Kelly, I. (1966). Ecological constraints on mental health services. *American Psychologist*, 21, 535-539.

Mills, R. & Kelly, J. (1972). Cultural and social adaptations to change: A case example and critique. In *Handbook of Community Mental Health*. New York: Appleton-Century Crofts.

Weiner, B. (1980). *Human Motivation*, New York: Holt.

9

Psychology and Social Problems

It is a matter of common knowledge that the contemporary Indian society is afflicted with innumerable social problems, such as widespread poverty, deprivation, illiteracy, child abuse, communal disharmony, rampant corruption in governance, increasing incidents of rape and abduction, demand of dowry, bride bashing and burning, casteism and communalism in politics etc. Psychologists in India realized it long back that these problems give rise to varied stumbling blocks in smooth functioning of the democratic form of governance and social development. Naturally enough, a good number of social psychologists in this country directed their efforts, either individually or in teams on studying some of such problems with a view to finding out scientific answers to three basic questions: (1) What are the manifest features and deleterious effects of the chosen problems? (2) How does the social problem manifests itself in what frequency that it disturbs peaceful and progressive social life? and (3) What are the specific sources and causes of the targeted social problems?

Since mid-fifties of the last century a number of social psychologists have investigated, and the work is still in progress on various types of social problems. During the past four decades numerous research articles, monographs and edited books have been published. A critical review of these publications reveals a number of meta-theoretical biases, blind spots, and neglect of well documented facts and facets of social reality in planning, execution and drawing conclusions from the data obtained.

This chapter tries to point out those biases and blind spots that made psychological studies of chosen social problems ineffective in providing sound scientific understanding of the origins and the causal factors of a single social problem. Thus, it is no wonder that no effective and dependable intervention programme or technique has been developed for solving a single social problem in any part of the Indian territory.

Bias of Environmentalism

The first half of the twentieth century was an era of different schools in psychology. Studies on mental processes and behaviours were conducted within the diverse frameworks of meta-theoretical assumptions. The most prominent meta-theoretical perspective was initiated by J.B. Watson. He forcefully argued that psychology as a science must focus on objectively observable behaviours. He emphasized that all behaviours are caused by stimulations received from the environment in which organisms are born, brought up and have their ambience. Environmental features were accepted as the determinants of all behaviours, Watson (1926; p. 10) stated that *'give me a dozen healthy infants, well formed, and own specified world to bring them up in, and I will guarantee to take anyone of them at random, and train him to become any type of specialist I might select—a doctor, lawyer, artist, merchant-chief, and yes, even into beggar man and thief, regardless of his talents, penchant, abilities, vocations and race of his ancestors'*.

The Watsonian environmentalism was further enlarged and promoted by Hull (1943, 1951, 1952), Mowrer (1947), Skinner (1938, 1953) and a host of others under the rubric of 'learning paradigm'. They argued that all living organisms including human beings learn various kinds of habits, habit family hierarchies through stimulus-response associations. This kind of behaviouristic school had become the most dominant and popular approach among psychologists of USA and some other countries as well as in India. It is worthwhile to note that it was in fifties and sixties of the twentieth century that a large number of young scholars and teachers of psychology went to USA for their doctoral degree of advanced research on various scholarships and fellowships. Due to rapid and wide expansion of psychology in Indian Universities, these psychologists made environmentalism as one of the most popular meta-theoretic approaches. This trend is still continuing in a large number of psychology departments. During the last four decades of the last century newer developments in the mainstream of psychology, the influence of environmentalism and the learning paradigm waned. However, it continued to dominate and shape social psychological researches in Indian Universities.

At this point it is worthwhile to note that the environmentalism and learning paradigm dwindled and was left behind, and some of its insights were incorporated in the mainstream cognitive psychology. For a fairly large number of psychologists from all over the world, environmentalism had a fresh incarnation in the corpus of cross-cultural, cultural and indigenous psychologies during the last three decades of the twentieth century. A fairly good number of psychologists assumed that culture constitutes the man-made environment, and it determines human behaviours and cognitions. However, when the question of providing a substantive definition of culture was raised, there were as many definitions as there were cross-cultural psychologists. Culture is conceptualized as 'some force out there to be studied' (Segall *et al.*, 1992) and 'as an all encompassing context' (Pandey *et al.*, 1996). Some psychologists conceptualize culture as a medium in which individuals grow psychologically. Such conceptualizations are rephrased form of environmentalism.

Any context, be it cultural or eco-cultural, assumed as the main determinant of the origin and growth of psychological functions is misleading. Such accounts remain limited to phenotypical aspects of psychological functioning. The genotypical facets of behavioural origin and development are either deliberately or unconsciously left out. Validity of such an assertion becomes self-evident, if and when one considers the methodology employed by environmentalists or cross-cultural psychologists. These researchers draw two samples from two highly different kinds of context, measure some features of mental or behavioural characteristics, and explore differences between the two sets of data. However, they ignore within group differences as error variance. It is in this error variance that spontaneously manifested genotypical effects are present but are ignored.

To illustrate the above point, let us take an example from agricultural sciences. An orchard man decides to plant saplings of four or five types of mangoes in the same kind of fertile plot. He provides same kind and quantity of irrigation, fertilizers and pesticides on all the plants. After a few years the plants become trees and yield fruits. However, *prima facie* all the fruits are more or less similar in shape and tints. But different varieties of mango seeds bear fruits of different sizes and taste which are easily discernible in spite of the fact that the mango seeds grew and came to bear fruits in the same kind of environmental background features. Social psychologists all over the world in general and India in particular have shown blindness for the realities and importance of the role of seeds (biogenetical contributions) in human development.

Neglect of Biogenetic Perspective

More than three decades back a sizable number of social psychologists came to realize the importance of the role genetic factors play in psychological development of every individual. Jensen (1968) asserted that environmental determinants of human behaviour and psychological traits do not provide a balanced picture of social behaviour. Biological inheritance as a source of psychological traits is of vital importance. The nature-nurture controversy arose a long time ago, and passed through a variety of vicissitudes owing to faulty questions raised by those, who were endeavouring to resolve the distinctive roles genetic factors and environmental variables played, in psychological development of human beings. Even- though they were convinced that genetic potentials along with environmental factors jointly determine human development, they raised wrong questions in the context of intellective functions. It is as well-accepted dictum that wrong questions never yield right answers. Two questions were mainly addressed to, i.e. which is more important in psychological development, environment of birth, existence and growth, or genetic potentials inherited from their parents and ancestors? The second question was about relative contribution of genetic potentials and environmental variables. After a prolonged debate and research efforts for providing valid answers it was realized that both genetic potentials and environmental variables are absolutely essential for human organisms to come into existence, survival and development. Psychological development is thus joint function of genetic potentials and environmental components. Both are singly necessary and jointly sufficient.

Psychological trait variability is a two-facet function, i.e. genotypical and phenotypical. The term genotype refers to genetic constitution of an individual, meaning thereby genetic base of a human being. It may be considered as inherited potential for specified traits, behavioural features observed in human beings. On the other hand, phenotype refers to observable outward characteristics resulting from simultaneous influence of genotype and environmental conditions. It is worthnoting that the stronger the genotypic potential, the more pronounced is its manifestations in the form of phenotypical characteristics. It happens to be so even if the person is born and brought up in highly impoverished environment. On the other hand, the weakers are the genetic potentials, the phenotypical characteristics remain weak eventhough the person concerned is born and reared in enriched environment.

Murphy and Kovach (1972) have elaborately discussed the relationship between genetic potentials and behaviour development. They conclude that 'the gradual disappearance of the classic nature-nurture problem, and its replacement by various complex issues of interdependence between organism and environment required more sophisticated use of behavioural analysis....' In the light of numerous research findings, Hirsch (1969) called for a complete reorientation of psychology. It has to follow the route charted by life sciences and has to accept its legitimate roots. At present widespread realization of complex and fine interplay between unfolding genetic potentials and individual experiences is at the centre of newly emerging biogenetic outlook in psychology.

At this point it is pertinent to note that studies of schizophrenics have shown that blood relations of schizophrenics are more prone to become schizophrenics than are random controls even when detailed exclusion of environmental effects are achieved. It has also been empirically shown that the chances of becoming a psychopathological case remain great even if the offspring from pathologically affected parents is removed (Kety, 1970).

A number of studies in the area of psychogenetic and genetics of behaviour development had been reported in which genetic manipulation of some major behaviours such as learning, drive and emotion were emphasized. Other studies laid emphasis on strain and genotype dependent differences in the process of behavioural development. Such studies have strongly proven that genetic factors are of vital importance in behavioural characteristics and their origin and development.

In an article on 'Code of Behaviour' in *Times of India*, it was stated that scientists of the Institute of Psychiatry, London, have been scanning human genome for answers for the behavioural question. According to them, most behavioural traits are results of interaction between nature and nurture. They have concluded that bad behaviour free of psychopathy had relatively little genetic component whereas psychopathic traits seem to be gene related. This finding is significant. But, if socially unacceptable behaviour is programmed into a person's genes, then would a conscious effort to modify behaviour prove effective? The findings of London scientists suggest that inherited criminal or psychopathic behaviour could be attended to with medical intervention only. Tweaking genes should not be seen as an answer in dealing with aberrant behaviours, because inheriting a tainted gene is not the same as the disease or disorder manifesting itself. It is well known that psychopaths are those who have different shades of criminal and unethical tendencies even though they are generally socially pleasant, amiable and not easily detected. Such psychopaths are responsible for initiating social problems, and their perpetuation is mostly due to environmental precipitators and modelling.

With a view to illustrate the preceding account of the role of genetic potentials in psychological development let us take up an example of a social problem and analyze it. It is a matter of common knowledge that fairly frequent communal tensions and clashes occur in India. One community organizes a religious procession, which passes through an area in which mostly members of another community have their residences, shops, and places of worship. The processionists shout slogans; use noisy musical instruments and loud speakers in large size vehicles. While the procession moves on through the areas of the other community, some residents of that area raise objections against the clamorous activity because it appears to be offensive. Gradually it turns into a confrontation between the two communities. While verbal exchanges are going on, the whole atmosphere hots up. Suddenly some members use firearms from both sides. The crowd created by the two gets transformed into mobs a good number of members of the two sides become de-individuated and violent behaviour occurs. Some persons are shot at, causing serious injuries, shops are looted, doors are broken and put on fire. After a while the police force intervenes, violence is suppressed, and tension filled calm is restored.

If one analyses the constituents of the two communities, it would become self-evident that all individuals in the two groups are not fanatic members of their communities. Some being orthodox members are attired in clothes indicative of community customs. Such members become aggressively vocal and adamant about their action plans. They may be considered as the front-liners and leaders. A sizeable number of participating persons in the two groups are followers of their leaders and conform to what their leaders dictate and suggest. The third set of participants in the two groups, who are there as community members with superficial interest in the matter and follow others without getting involved in the issue. Another set of persons in the two community groups do not feel comfortable in this kind of tension and possibility of violence. They are there because they do not want to be isolated in the community. When the situation becomes violent, some individuals stealthily walk or run away. However, some members of both the groups, predisposed to getting involved owing to their bad genetic potentials for illegal/criminal predispositions, perceive the situation fit for manifestation of their latent psychopathic traits. Such persons would be present at any layer of the crowd, or it is this fringe that initiates illegal/criminal disruptive actions. Once such activities begin, the crowd turns into mobs. Now individuals become de-individuated and there is a large-scale use of firearms, shop looting, property burning etc.

Proper understanding of different kinds of social problems warrants thorough analysis of various layers involved in them.

Such analysis is bound to differentiate the role of environmental features and genetic factors. Apart from this, in order to develop effective social technology for solving problems, it is essential to determine the level of human evolution of people who get beset by social problems.

Importance of Human Evolution

Theory of biological evolution deeply influences the sciences of anthropology, sociology and economics. Owing to this influence, studies of social, cultural and economic activities came close to the interest of psychologists in general and social psychologists in particular. Now it is a matter of common awareness that nowhere in the world social life remains static. Easily perceptible changes occur from one generation to the other in the forms of social rites and rituals, traditions, customs and relations. Changes in the substantive aspects of social activities and thinking occur relatively more slowly and lead to onset of a different state of human evolution. The six volumes of August Comte (1830-42) on Systems of Positive Polity became a source of profound influence in theory of human evolution. According to him there are three stages of human evolution, i.e. (1) theological, (2) metaphysical and (3) positive or scientific. It is a rather broad classification.

Theological Stage

This stage began during that period when human beings became capable of simple perceptual discriminations, retention, storage and retrieval functions of memory and constructive imagination. According to Taylor (1871), the onset of this stage was characterized by animism. In this context he presented a vivid account of primitive mentality. Primitive man, according to him, thinks of this world as a place of animated beings existing on this terrestrial sphere. They are either inimical or friendly to man. In the other natural forces, products and objects are either benevolent or malevolent. Rivers, clouds, mountains, trees, yielding eatables, timber or firewood are considered benevolent, while snakes, carnivorous animals, poisonous plants are considered to be malevolent.

Primitive man makes a distinction between body and soul. The former being mortal and perishable, while the soul is immortal and non-perishable. It is in this primitive mentality that a religion originates. All benevolent entities are worshipped for their blessings. On the other hand all malevolent ones are either, to be avoided or to be approached and offered suitable sacrifices. Gradually such religious practices give rise to different kinds of theologies.

All theologies are rooted in tenacious faith, rituals and rites which are prescribed and promoted by priests, clergymen, pundits, purohits, imams, mosque keepers and rabbi for whom their work is a vocation providing means of decent livelihood. In a theology there is little scope for dissent. Dissenters are ostracized and often tortured. Theological principles along with prescribed rites and rituals are shaped by ecological conditions of the territory in which a particular religion thrives. In all theological theories and assumptions on the omnipotent and omnipresent, God is conceptualized and accepted as the supreme power that controls the universe and rules over the destiny of human beings.

Metaphysical Stage

As there is an increase in cerebral activities, thinking capacity and reasoning power by virtue of biological evolution, some individuals find contradictions in untenable assumptions in the theologies concerned. Such individuals raise their voice against the controlling power of the guardians of theology; they propound their self-created theory of life, time, space, identity, nature etc. Thus newer philosophies are created with diverse kind of principles and precepts, some woven in theism and others in atheism. It must be noted that all types of metaphysics are rooted in *a priori* reasoning and thinking. The quantum of metaphysical thinking increases, the theological practices undergo drastic changes. In each religion diverse types of sects and sectarian programmes are launched. History provides authentic evidence to prove that well educated individuals, developed in the prevailing system of education endeavour to bring about changes in the style of life for the masses. Some of those entrenched in certain metaphysical ways of life do not feel fully satisfied and become restless for more dependable understanding of life on earth. This kind of restlessness paves the way for moving towards the positivist or scientific way of understanding and knowing.

The Positive or Scientific Stage

This is the final stage of human evolution. This stage is rooted in the empirical methods. During this stage social reconstruction is affected through emancipation of mankind from *a priori* reasoning based bondage of metaphysics in favour of the practice of direct and objective experience of natural phenomena. So acquired insights are available for experience to the public. Scientifically tempered persons neither accept nor reject any statement about natural entities. They put a question mark against it. If and when their interest is aroused they examine such statements on the touchstone of objective or controlled observations. In this stage, quite often efforts are made to identify and define specific antecedent conditions and note down their consequent outcomes. This is done to discover the laws and principles of natural features. Scientifically oriented researchers assume that *Nature* functions according to specific laws. If these laws are correctly understood, they not only explain but also lead to their control and predictions. However, this positive stage is in its infancy and it is still to go a long way in order to understand numerous mysteries of Nature.

It is important to note that there is no clear-cut transition from one stage to another. At present all the three stages exist simultaneously all over the world even though they originate in a sequence. It is so, because human evolution depends on one's genetically endowed intellectual ability, environmental enrichment or impoverishment, and upbringing including education. All over the world there are communities and social groups, which are still living, in the theological stage. Lots of communities educated in traditional schools and having spiritual interests are in metaphysical stages and are following saints, swamis and scholars of scriptures. A fairly sizeable number of individuals have entered into the scientific stage.

Origin and Solutions of Social Problems

Before an effective intervention programme or remedial measure or social technique is developed, it is absolutely necessary to have scientific answers to three basic questions, i.e. (1) what are the salient features of social problems and specific behaviours that characterize the problem? (2) what is the level of human evolution of those who add to the problems and those who gets affected by it? and (3) what kind of upbringing, socialization, the initiators and perpetuators of the problem have, as well as, what kind of ambience they live in? Answers to these questions are necessary requirement for developing an effective remedial technique.

With a view to finding reliable and valid answers to these questions so that an effective technique is developed, is a stupendous task. It requires a long-term plan to be implemented by a team of social psychologists. Such plans should have several steps to be sequentially implemented. First, the problem-afflicted segment of society must be identified and delimited. The second step should be identification of the initiators and perpetuators of the problem. The third step calls for development and standardization of effective testing techniques for ascertaining the relative contribution of environmental background including upbringing, schooling and areas of ambience, family history including ancestral antecedents, and level of psychological development. In the light of research findings reported so far it is reasonable to assume that the relative contribution of these factors would vary from problem to problem as well as from person to person in society.

How to determine the role of these factors is a complex task, and a single researcher, no matter how competent and hard working he is, it would not be possible for him to complete it in his lifetime. Such a task calls for a teamwork. Here, what is warranted is that various councils, commissions and organizations established by the Central/State Government should take interest and take initiative. They should launch all India level seminars on specific problems and research projects. This task should be the main theme of deliberations in the annual conferences of the professional organizations of psychologists, sociologists and anthropologists. The governments on their own cannot solve social problems by enacting laws and trying to implement them.

In this context the next but the most important task is concerned with development of effective remedial techniques for solving different types of social problems. Notwithstanding the fact that a number of psychologists and other social scientists have studied a variety of social problems in India, and have endeavoured to develop intervention programmes, no effective and acceptable social technique has been prescribed. In an article entitled, 'Is the study of society science?' Alan Ryan states that the most familiar implication of the absence of general laws on par with natural science is that there is nothing one could properly call social technology.

Whatever techniques are available for social engineering are not properly utilized. A fairly large segment of Indian society is against the use of such technologies because it is considered to be a violation of the freedom of choice. It is argued that to treat other people's behaviours, beliefs and practices as objects for technical manipulation would amount to transgression on God's domain, and violation of the democratic principle of the freedom of choice. However, what is valued as freedom of choice is nothing more than leaving every improvement to random and ill-understood programming by the environment. Now the relevant question is: would it not be better for everyone that the programming is made rational, controlled and the effort is fully aimed at maximizing mental peace, social harmony and happiness. At present what is needed is that the social psychologists, sociologists, anthropologists and psychiatrists get together and concentrate on some selected social problems with interdisciplinary framework and discover the causal factors, and develop some effective techniques to solve that problem. Medical sciences have done so in the areas of prevention and betterment of physical health.

References

Compte, Auguste (1830-42). *Cours De Philosophies Positive* [System of Positive Polity] 6 Vols. Paris: Bacheller.

Hirsch, J. (1967). *Behaviour-Genetic Analysis* (ed) New York: McGraw-Hill.

Hull, C.I. (1943). *Principles of Behaviour*. New York: Appleton Century-Crofts.

Hull, C.I. (1951). *Essentials of Behaviour*. New Haven, C.T.: Yale University Press.

Hull, C.I. (1952). *A Behaviour System*. New Haven, C.T.: Yale University Press.

Jensen, Arthur (1968). Biogenetic Perspective: Introduction in Martin Deutsch, Irwin Katz, Arthur & Jensen (eds) *Social Class, Race and Psychological Development*.

Kety, S. (1970). Genetic Environmental Interaction in the Schizophrenic Syndrome. In R. Conero (ed) *The Schizophrenic Reaction*. New York: Brunner-Mazel.

Mowrer, C.H. (1947). On the dual nature of learning: A re-interpretation of conditioning and problem solving. *Harvard Educational Review, 17*, 102-148.

Murphy, Gardner, & Knovach, Joseph K. (1972) *Historical Introduction to Modern Psychology* [sixth edition] p. 366. London, Routledge & Kegan Paul Ltd.

Pandey, J., Sinha, D. & Bhawuk, D.P.S. (Eds) *Asian Contribution to Cross-Cultural Psychology*. New Delhi: Sage Publications.

Seagull, Marshal, Lonner, J. & Berry, J.W. (1998). Cross-Cultural Psychology: A flowering Discripline. *American Psychologist, 55*, 1101-1110.

Skinner, B.F. (1938). *The Behaviour of Organisms*. New York: MacMillan.

Taylor, E.B. (1871). *Primitive Culture*. London: Murray.

Times of India (Lucknow Edition) Code of Behaviour. June 11, 2005, p. 9.

Watson, J.B. (1926). *Experimental Studies on the growth of emotions*. In C. Murchison (Ed.) *Psychologies of 1925*. Worcester, M.A.: Clark University Press.

Part II

Teaching Profession

10

Some Motivational Aspects of University Teachers

The present chapter is rooted in cumulative impressions, casual but close observation of academic activities and behaviours on a variety of campuses, frequent informal discussions with in-service or retired colleagues inside as well as outside various university authorities, and also in the data culled from unsystematic content analyses of news, statements, editorials and topical write-ups about higher education and university life. The assertions made in this paper, therefore, should be taken as hypothetical propositions, and in no case as empirically verified facts or conclusions. The analysis of motivational variables underlying the behaviours and activities of contemporary university teachers is articulated here in terms of the mastery and growth theories of humanistic psychology of Maslow (1970) and also in terms of insights gained from a plethora of researches on interactions between intrinsic *versus* extrinsic motives and rewards (Brehm 1966; Deci, 1975; De Charm, Jones *et al.* 1972; Kelley, 1967).

It must also be made clear at the very outset that the present paper is an endeavour to analyse and evaluate the motivational components of the contemporary academic and extra-academic activities in university campuses with reference to the motivational properties to be found in an idealized university teacher. These properties are to be kept in mind as standards against which the motivational properties underlying the behaviours of contemporary dons on the campuses are to be assessed. Validity of the assertions made in this paper depends on the acceptability of the concept of an idealized teacher.

Motivation of an Idealized University Teacher

Like any other normally developed, educated, and healthy human being a university teacher must have full-blown need for a stable, firmly based, unusually high evaluation of himself, for self-respect and also a strong need to realise his potentials in the processes underlying acquisition, dissemination and extension of knowledge and skill in the chosen discipline. An idealized teacher must possess intense desire and indefatigable zeal for intellective strength, for academic achievement, for adequacy, for mastery and competence, and for self-confidence and independence. Apart from these needs, he is also expected to have need for respect or esteem from other people, need for dominance, recognition, attention, importance, dignity and appreciation. On the other hand it is also expected that an idealized teacher must be capable of guarding himself from the danger of basing self-esteem on the opinions of other rather on his real capacity, competence and adequacy to the task. His need for self-esteem should find gratification in deserved respect from others rather than on external fame or celebrity, and no unmerited adulation. In brief, such a teacher must have intrinsic motivation for academic activities. His survival or homeostatic needs and motives should be subservient to his need for realizing his potentials as a seeker and sharer of knowledge and skills.

Such a conceptualization of an idealized teacher must not lead to the faulty impression that he should be completely either gratified in, or devoid of, the needs for physiological homeostasis, for security and safety and for love and belongingness. What is expected of him is that such a teacher must have prepotent intrinsic motivation for academic pursuit, and must also have the proclivity for depending on his internal resources, ability and effort in satisfying his needs rather than on external sources, powerful others and manipulative skills. By any stretch of imagination it must not be conjectured that an idealized teacher is an ethereal being living in a world of ideas and ideals only. What is expected of him is that he must not become compulsively addicted to materialistic pursuits but must become a truly ardent worshipper of the goddess of learning, knowledge and self-actualization.

Present Day Motives of University Teachers

During recent times in this country the number of universities and university teachers has increased by leaps and bounds. In the backdrop of the national goals, aspirations and activities the university teaching community as a whole has developed certain motivational characteristics which are mostly discrepant and opposed to the motivational characteristics necessary for making a university teacher move in the direction of becoming an idealized teacher. In my view, a large number of university teachers has unwittingly developed motives that have contributed a great deal to the overall deterioration of the academic health on campuses.

Assertion 1: Among the contemporary university teaching communities extrinsic motives and goals have become prepotent and led to weakening of the intrinsically motivated activities and gratifications.

Such an assertion in unequivocally supported by a variety of phenomena that are easily discernible in the activities on the campuses. Proliferation of extra remunerative assignments and consequent rat-race among a large number of teachers for obtaining these assignments, publications of cheap text books and sure-success-in-examination guides, ill-conceived and sloppily written research papers to be published in commercial journals, ever-increasing number of university teachers taking up private tutions and running all kinds of side-business activities, hectic activities for procuring extra remunerative examination and admission test assignments, trade union like group efforts for exclusive purposes of negotiating with the governments for the triple benefit of provident fund, pension and gratuity, for rotational systems of headship and other economic and material benefits etc. have become conspicuous. Such activities have to be considered as having vitiating influence on genuine teaching and research. Enhanced preoccupation with pursuit of extrinsic goals not only distracts the university dons from the vital course of academic activities but also attenuates the gratification of potentialities of the intrinsically motivating activities.

Assertion 2: (a) *The prevalent national emphasis on economic and industrial development of the country has led to unprecedented increase in the extent and magnitude of physiological deficiency and comfort needs, safety and security needs, and needs for love and belongingness among university teachers.*

(b) *On the other hand such augmentation of deficiency needs, goals and values among the university intelligentsia has kept its large segments preoccupied with these needs and has also retarded the emergence and proper development of the growth needs such as need for self-esteem and self-actualization.*

Implications of these two interrelated assertions become self-evident if one examines the present day activities and living patterns of university teachers. It is no secret that during modern times the motto of simple living and high thinking has been rejected and replaced by affluent living and light reading or television watching. More and more teaches are preoccupied with building houses, buying and enjoying modern electrical and electronic gadgets, furnitures and furnishings for comfort, making efforts at savings for future use, and also busy with buildings and supporting unions and participating in union activities with a view to become persons of enhanced pressure value in order to derive increasing degree of concessions and material benefits from the government and other statutory authorities. On the other hand, it would not only be interesting but highly revealing to find out what percentage of teachers spend what percentage of their income in purchasing books, magazines and journals for serious reading and what proportion of time they spend in reading and doing other academic work.

If you take a close look at the deed, diction and deportment of a representative samples of university teachers you can easily guage the extent to which they manifest the behavioural referents of self-esteem needs. Here, instead of presenting phenomenal descriptions with a view to illustrating either complete lack of the emergence of need for self-esteem or its deviant course accompanied by feeling of inferiority, weakness or helplessness giving rise to basic discouragement or compensatory, and not infrequently to neurotic trends in activities on campuses. I would rather raise some pertinent questions in this respect and expect readers to ruminate over them. Is it true that during recent times a fairly sizeable number of university teachers indulge in favouritism, carelessness, in persuit of vested interest and non-specifiable underhand dealings in respect of admission and examination work? Are there teachers who have become indifferent to ethical standards in their teaching and research assignments? Is it not a fact that there are many teachers who now consider it proper and necessary to pull political strings and indulge in ingratiation techniques in relation to powerful others with a view to scurrying favours in matters of promotion and getting important assignments? If your answer to these questions are positive, the aforesaid assertion intuitively becomes valid.

Assertion 3: *A sizeable number of teachers has normal and healthy motivational development and their self-esteem need potentials are undoubtedly prepotent. However a fairly large number of such teachers, exhibit compensatory or deviant behavioural trends and endeavour to gratify their needs for status, fame, recognition and dignity not through genuine achievement of professional adequacy and competence, and also not through mastery over the chosen discipline of knowledge, but through the use of manipulative skills and ingratiation strategies with a view to procuring the externally visible signs of status and recognition.*

The veracity of this lengthy assertion becomes self-evident if one takes a close look at the under-the-surface activities of university teachers inside as well as outside the campuses. On some campuses what kinds of consideration dominate in the matters of appointments and promotions, would be revealing to find out. Are there no teachers who spend their prime of time in building political connections, in ingratiating themselves before persons occupying positions of power? How are nominations to various national or regional or university academic bodies and committees made some time? Is it rare to find teachers who are more like political workers? Is it uncommon to come across award of doctoral degrees, and acceptance of research projects on all kinds of consideration but purely academic? Is it conceivable to have satisfactory explanations of deterioration in higher educational standards without assuming detractive role played by some teachers in the educational processes? Can we say that positions of power and prestige in universities are always assigned to deserving persons and that too on genuine considerations of academic excellence? I leave these questions to readers.

Assertion 4: *During recent times the tendency to be externally controlled has increased and the essential characteristic of remaining internally controlled has declined on one hand, and the tendency to attribute success to one's ability and effort, but failures to interfering powerful others as well as to vagaries of circumstances and chance has phenomenally increased on the other.*

This assertion is too obvious to require elaborate explanation and illustration. Are we not aware that a large section of the university teaching community regards the Vice-Chancellors and members of the selection committees, functionaries of University Grants Commission, Boards of Higher Education, ICSSR, CSIR and other such bodies and politicians-in-power as the controlling agents of rewards and punishment? Is it not a fact that it has become a common practice to exclusively attribute all ills, that are corroding the very vitals of academic processes, to external factors such as rampant student unrest and indiscipline, to political interference, to public apathy and to degenerated norms of pubic behaviour and to declining ethical standards in politics and society? Is it not a common practice that quite a few university teachers attribute their successes to their ability and effort even though they have achieved them through dubious means, and for their failures blame others and impropitious circumstances? Have not the internal factors contributed to the prevailing ills in the current academic processes? Do university teachers not owe it to themselves to introspect and assess their active roles in the prevailing academic ethos?

So far the present chapter appears to project an extremely gloomy and pessimistic view of the motivational aspects of university teachers. However, it is not a complete picture. Amidst these deviant motivational characteristics, an analysis of a small minority of university teachers yields some bright spots, which ought to be described.

Assertion 5: *Though in absolute terms increasing, but in relative terms, decreasing number of teachers in modern times has, against all odds, developed beyond the prepotency of extrinsic motives, deficiency needs and values, and is adequately motivated by self-esteem needs of mastery and competence, as well as of reputation and recognition. Such teachers are predominantly subject to internal control and adopt objective and rational approach in their attributional analyses of causal factors underlying successes and failures.*

If one compares the university teachers of yore with those of the modern times there is still a large number of teachers who have healthier motivational force behind their professional performance and academic contributions. It is an indisputable fact that in spite of vitiated academic atmosphere, standards of research, of creative writing, of effective teaching and other academic contributions in some departments of some universities located in various parts of the country have gone up. But the number of teachers setting up such examples of academic excellence and exemplary motivational patterns constitutes rather a microscopic minority in the community of university dons.

In the end, I must admit that these assertions about the motivational characteristics of university teachers may be equally applicable to other professional communities. Such a generalization may be true, but is irrelevant in the present context. As a matter of fact such a comment should be construed as an attempt at detraction from the crux of the problem with which the university teachers in general are confronted. These assertions have to be thoroughly examined, both empirically and rationally, and ways and means will have to be found out to promote correct kind of motivational patterns among the present day and future university teachers.

References

Brehm, J.N. (Ed.) (1966). *A Theory of Psychological Reactance*. New York: Academic Press.

De Charms, R. (1976). *Enhancing Motivational Change in the Class-rooms.* New York: Inengon Publishers.

Deci, E.D. (1975). *Intrinsic Motivation*. New York: Plenum Press.

Jones, E.E., Kanouse, D.C., Kelley, H.H., Nisbett, R.E., Valins, S., & Weiner, B. (Eds.) (1972). *Perceiving the Causes of Behavior*. Morriston, N.G.: General Learning Press.

Kelley, H.H. (1967). Attribution theory in social psychology. In D. Levine (Ed.) *Nebraska Symposium on Motivation*, University of Nebraska Press.

Maslow, Abraham (1970). *Motivation and Personality*. New York: Harper & Row.

11

On Reward and Punishment in Indian Universities

No one can deny the fact that today, in India, higher education is not what it used to be four decades back. Apart from the fact that during the past forty years there has been mushroom growth in the number of universities, colleges and other institutions of higher education; drastic systemic changes have been introduced in the structure and organization of universities that have culminated in functional changes of far reaching consequences. Internal control and autonomy have been replaced by external surveillance and sanctions. Self-imposed discipline, self-initiated and self-sustained commitment to acquisition and dissemination of knowledge, which used to be the hallmark of university life, have been gradually relegated to background by statutes-defined rules, length of service and rotational representation in decision making bodies. Emphasis has shifted from learning and teaching to examinations, award of degrees and professional designations. The concepts of national character, academic accomplishments and professional involvement have now become overshadowed by regional identity and character, seniority and length of service, extra-academic capacity and resourcefulness. In running the academic, financial and administrative affairs of the universities democratic process and procedures have gradually been undermined and replaced by authoritarian decisions and bureaucratic impositions. And, finally, the repositories of intrinsic motivation, self-reinforcement and assiduous pursuit of competence have now become hot houses of extrinsic restlessness, incentives, and monetary rewards. It is this change, that is most tragic and despairing.

Some Fundamentals of Reward and Punishment

Let us look at the reward and punishment conditions prevalent at present in the realm of higher education. It is needless to point out that higher education demands attainment of excellence in academic or intellectual activities from the students and teachers alike. Such activities are best performed if they are intrinsically motivated, meaning thereby that, these activities have to be rewarding and pleasurable in themselves, and create a feeling of continuous self-enhancement. Such activities are further butteressed if they bring recognition of competence, upward movement in status and designation. However, in order to attain such excellence one has to follow the principle of best effort which runs counter to the inherent biological tendency of the principle of least effort. Naturally, in any institution if the observance of the principle of least effort becomes socio-academically rewarding, the principle of best effort will gradually become weakened and ultimately extinguished. We shall shortly see how inadvertently or deliberately various activities in higher education are getting so geared that the principle of least effort is gradually becoming more rewarding and the principle of best effort is indirectly becoming more punishing and frustrating.

While scanning the affairs of contemporary higher education, we have to keep in focus the concept of vicarious reinforcement. One does not have to be personally rewarded or punished for strengthening or weakening one's action tendencies. It is now a commonly accepted principle that one learns a great deal through observation of activity-reward and activity-punishment contingencies in one's surroundings; and the observer's behaviours are strengthened or weakened vicariously through observation of others who are rewarded or punished for such behaviours.

One more thing about the fundamentals of reward and punishment before we take a bird's eyeview of higher education in most of the Indian universities. Any event or object initiated as reward or punishment becomes rewarding or punishing in the context of the motive-incentive conditions of the person or group concerned. Ordinarily no event can be considered as rewarding or punishing unless and until the person has appetite or aversion for that event in a specific context. Conferment of recognition to a money-motivated person is not only non-rewarding but at times it may prove to be punishing and frustrating. Perhaps, barring money, no object or event functions as reward independent of the motivational dispositions of the person in the context. Thus, reward and punishments are subtle functions and have to be skilfully employed for desired effects. However, as we shall shortly see, in most of the Indian universities, systemic changes have led to such reward and punishment structures that extrinsically motivated activities are more often than not rewarded, and at times intrinsically motivated academic and scholastic creativities and attainments are directly or indirectly punished, or, at best, ignored. Over the past few years such functioning has led to fast corrosion of intellectual life on the campuses. As a matter of fact I very strongly feel that the inherent human proclivity for acceptance of the principle of least effort has vastly suppressed the necessity for adopting and observing the principle of best effort without which scholastic activities cannot be generated and sustained. It seems that the institutional climate of campuses is fast becoming more congenial for the growth of extrinsic needs for material well being and social recognition and more sterile for intrinsic need for understanding, creativity and self-enhancement. Almost all over the country, one only has to look behind the apparent life of university campuses, and one comes across glaring illustrations of this tendency.

Institutions of higher education are established, developed and maintained for acquisition and dissemination of knowledge and skill with emphasis on their social relevance, so that these institutions may function as moulders of required human resource for all-round national development. Keeping this function of the institutions of higher education in focus, one may examine the contemporary characteristics of the various segments of higher educational institutions and assess their contributions within the framework of their reward-punishment functions.

Merit Promotion Scheme

How has the inclusion of extrinsic motivation brought in gradual corrosion of the very vitals of academic standards and activities can be illustrated by the merit promotion scheme which came in force a few years back. It is common knowledge that prior to the introduction of this scheme the designational hierarchy in Indian universities was pyramid shaped, and ascendance to the top of the pyramid was contingent primarily on academic merit and professional excellence instead of length of service alone. Even though long in the length of service, but academically stagnant teachers had extremely slender chance of being elevated to the positions of Readers and Professors. Unless and until one kept academically growing and enhancing one's excellence in acquisition and dissemination of knowledge through publishing research papers, monographs and books, and establishing oneself as noted academician, one earned bad reputation and got neither recognition nor promotion. All higher posts were filled through open selection making migration of teachers from one university to the other a common occurrence. It is also true that owing to existence of fewer posts at the top and much more posts at the bottom sometimes really meritorious people had to languish in lower posts. This defect would have been amended by creation of more higher posts in the universities. The University Grants Commission had found this solution and had initiated the scheme of sanctioning higher posts to universities in the past Five Year Plan periods. It is altogether a different matter that the conditions for filling up these posts were highly restrictive for state universities. However, appointments against those posts became such that instead of providing incentive to academic excellence, they highlighted the importance of extra scholastic qualifications and qualities. As a consequence such appointments provided an impetus to teachers for demanding promotional benefits.

University teachers from all over the country resorted to agitational approach, and the UGC ultimately accepted a merit promotion scheme to be implemented with some constraints. When the merit promotion scheme came to the State universities, the academic qualifications for promotion were so watered down that the whole scheme became virtually a time-bound promotion scheme, and a certain length of service became the only prerequisite for ascendance to higher posts with better emoluments and the resultant designations became the satisfiers of the need of recognition. With the recent introduction of the new pay scales, the futility of this scheme has been realized, and it has been abandoned. This scheme, however, has created precedents of vicious contingency in which one is surer of getting promotion through length of service rather than through acquisition of academic competence. The pervasive aftermath of this precedence-setting reward contingency will be difficult to avoid in future appointments.

Vice-Chancellors, Principals and Professors: Models and Directional Cues

In the organizations of higher education the top echelon consists of Vice-Chancellor or Principal, Professors and administrative officers. They function not only as sources of rewards and punishments, but more effectively as models of observational learning. Such learning is done by young teachers, senior students, ministerial staff and others directly under their charge. This type of learning is to a great extent incidental but crucial for all-round academic growth. The top echelon functions as beacons which the other members of such organizations endeavour to reach. If one scrutinizes the grounds and merits on which these functionaries are appointed these days, the kind of activities they indulge in, and the way they function, there emerges a highly disquietening picture. Barring a few universities and a few accidental cases, vice-chancellors are now appointed by the central or state governments on recommendation of committees which are largely the creations of the governments concerned. It is now an open, though not articulated, secret of the intelligentsia in higher education as to on what grounds, merits and personal qualities, persons are selected for vice-chancellorship. Whenever and wherever persons of questionable academic values and administrative abilities but of immense political clout become vice chancellors, they not only bring in administrative lapses and laxities, they also undermine or sidetrack high academic values and activities. Further too, they become models for all those who have the ambitions to become vice-chancellors. Naturally, if such persons happen to belong to the cadre of professors or principals, they are vicariously reinforced for attaining those non-academic merits and personal qualities which now are the frequently observed criteria of appointment as vice-chancellors, and consequently leading to detrimental effects on both academic activities and administrative efficiency.

In appointment of professors and principals, and more so, of other academic and administrative positions the criteria of selection are not always academic accomplishments, teaching capacity and research attainments. More often than not consideration of caste, community, region, political affiliation and recommendation etc. sway the balance against academic achievements of the candidates. Persons so appointed again become models for junior teachers and senior students, and they naturally endeavour to develop only those qualities which bring promotion, reward and satisfaction in life. When such appointments become common, the grounds on which they have been made also become implicitly acceptable at the social plane. Ultimately they assume the status of norms. After having had four decades of experience and interaction with a large faculty belonging to different disciplines in a large number of universities, I have gathered the impression that now extraneous considerations in appointments have become tacitly but boldly accepted norms though not yet admitted publicly.

University teachers and officers, so appointed, set rather lower standards of academic aspirations and activities, be they concerned with study, teaching and research or administrative efficiency and competence. Naturally students and others directly below their wings have lowered down goals to attain and remain underdeveloped. On the other hand, these teachers and officers, having been appointed on non-academic considerations, continue nurturing those qualities and considerations on bases of which they got their success. Naturally, they have neither time, nor inclination for doing what they are purportedly expected to do in the academic field.

Students of such teachers through observational learning emulate their behaviours and strategies for forging ahead in their chosen spheres of activities. These are the root causes of numerous malpractices that have become cancerous on the campuses.

Extra Remunerative Jobs

Since mid-seventies there has been phenomenal increase in the types and numbers of extra-academic jobs in the universities, and a fairly large number of university teachers are appointed against these part time posts which carry honoraria of one hundred to three hundred rupees per month and other fringe benefits, such as, free residential accommodation, conveyance, telephone and/or orderly-cum-peons.

Appointments against such part-time posts are usually made for three to six years, and appointees perform these jobs over and above their jobs as teachers. Some of these jobs are so time and energy consuming that these have to be performed at the cost of teaching and research.

Apart from these, there has recently been an unprecedented spurt in seasonal assignments requiring a few days to a few months of non-academic duties, such as supervision of admission or, selection tests and specially created jobs for teachers in regular university examinations. Then in many universities there are special extra-academic assignments created by vice-chancellors according to their whims. They are mostly given to teachers.

These seasonal or recurring extra-academic jobs have created a vicious circle of reward and punishment. If these jobs are assigned to teachers who are committed to pursue the path of academic excellence, either they do not accept them, or if they do, these jobs prove detrimental to their academic activities. These teachers are extrinsically rewarded for doing non-academic work. Consequently habits for extra-academic activities are strengthened and habits for intrinsically motivated scholastic pursuits are weakened. Precisely because of this reason quite a large number of truly committed teachers do not accept these distractive assignments. Naturally, more often than not teachers of conspicuously non-curricular attainments are given these assignments and thus rewarded for cultivation of non-academic qualities, and indirectly also for not being committed to what they had been appointed for as teachers. On the other hand, for junior teachers these appointees become models of monetary and social success, and they are motivated to, and vicariously reinforced for, following the course of non-academic activities and importance. Such motivations and vicarious reinforcements become more effective because they are in consonance with the principle of least effort. It is my belief that ever since there has been increase in extra remunerative jobs in the universities, academic activities, values and standards have taken a sharp downward trend.

Rewarded and Punished Behaviours of Students

Over the past few years I have tried to scan the academic climate in a large number of universities with reference to the quantum and quality of teaching and study, evaluation of academic achievement, teacher-student interactions, research activities and administrative processes in management of students' affairs. In this process of search a large number of startling and saddening facts have become evident. Barring a few exceptions, most of the universities especially in the north, have come in the grip of such socio-political climate that high standards of teaching and study are either neglected or ignored. Conscientious, objective and impartial evaluation of examination performance has fast withered away. Use of recommendations, exertion of social and political pressures and harnessing of a variety of non-academic means for getting good grades and marks including approval of theses for award of Ph.D. degree have become rewarding and rampant. Consequently, highly brilliant and industrious students do not always excel in examination results and thus are indirectly punished. The malady has gone so deep and its symptoms have assumed such alarming magnitude that all over the country, in both private and public sectors, or in government run and autonomous institutions, university examination results are no more considered as truly reflective of the degree holders' ability and scholastic achievement. Screening tests for selection of deserving candidates for appointment against different kinds of posts are now in vogue in almost all important areas of employment.

In the context of student activities let us look at reward and punishment in a more literal and mundane sense. Which types of students or what types of student activities are rewarded by allowing admission to hostels, award of freeship and other financial aids, enhanced library facilities etc.? Only an ostrich approach can prevent any intelligent observer of the present day university scene from discovering the extent to which groupism, political patronage and affiliation, muscle power and other non-academic (and at times anti-academic too) resourcefulness have become rewarding in a variety of university functions. Agitationist and activist groups of students manage to get the rewards of recognition, support and favours from not only university authorities but also from political parties in power or in opposition. In certain universities student union leaders are known to have secured unprecedented material and monetary benefits. It is now commonly accepted that to become politically active on the campus is more rewarding than to be academically industrious and accomplished.

Sometimes some rowdy or agitating students are punished in the form of being suspended from studentship pending enquiry of their conduct and character. Such punishments do not create noxious states for the punished, and therefore, seldom lead to suppression of undesirable behaviours. As a matter of fact, such suspensions and subsequent withdrawals create more pernicious effects on academic climate of campuses. Once in a blue moon, such suspensions are followed by expulsion and rustication from the campus. But the socio-political climate of the country is such that persons so punished use this punishment as exemplary political diploma and glory. The result is that suspensions and expulsions from campuses for politically motivated students have become more valuable and covetable than academic degrees and diplomas. In reality, most cases of such suspensions are subsequently withdrawn and provide political recognition to such students. In brief, the kind of punishment that is used on campuses, functions as valuable rewards or recognition, better status and greater clout with the university functionaries, instead of creating suppressive and deterring effects on anti-academic activities.

Oases in Fast Expanding Deserts

The above analysis of the academic climate prevailing in most of the Indian universities leads to the inevitable conclusion that misuse of rewards and punishments in academic functioning has gradually led to expansion of academic barrenness. However, even though everything may not be claimed to be in perfect order in the UGC, its functioning may be regarded as far from being commendable, and some of its plans and procedures may have dubious merits; the University Grants Commission, which is the main source of sustenance and developmental grants to Indian universities, has successfully created life saving oases by implementing the schemes of departmental support, special assistance, university leadership plans, fellowships, grants for research projects and textbook writing, national lectureship etc. Even though at the level of implementation whether all the aims and objectives of these schemes are achieved or not is questionable, it can be rightfully asserted that these schemes have initiated the desired kind of reward and punishment strategies for the growth of academic activities, values and accomplishments. There are other agencies also, such as ICSSR, CSIR, etc. which provide excellent reinforcers to those academicians who are truly committed to academic excellence. Of course, it can be forcefully argued that sometimes even these organizations have to yield to banal socio-political influences, but they still are engaged in providing the right kind of incentives, rewards and valuable directions to those who are still immune to perverting effects of present day maladies rampant in our socio-political ethos.

In the preceding paragraphs, only some conspicuous aspects of reward and punishment contingencies have been outlined. Today shrewdness, ingratiation with the establishment, lipservice to idealism have made love for wisdom, truth and knowledge, harmony in thought, speech and action, and endeavours for enhancing academic competence through study, research and intellectual discussions less rewarding and attractive.

12

Extra-Remunerative Assignments in the Universities

Once considered as temples of learning, a number of State and Central universities have now degenerated into dens of mammon. Focus of interest has gradually shifted from learning to earning, from simple living and high thinking to affluent life and pseudo-intellectualism, and from teaching to coaching. Teachers still work hard, but not as much for effective teaching and serious research as for extra-remunerative assignments. Most of the University teachers still write and publish but not all of them create original and scholarly treatises. Instead, a fairly good number of teachers write guide-books and 'sure-success-in-examination' notes. They still do research and some of them publish research papers and monographs, but more often than not, the driving force behind this endeavour for many of such teachers is not the motivation to extend the boundary of knowledge but to secure promotions and higher grades. The onset of this degradation has been precipitated, though predisposed by many factors, by proliferation of extra-remunerative tasks to be assigned to in-service teachers in a number of universities.

Background and Origins

With a view to opening the flood-gate of occupational and professional opportunities to the sons and daughters of all citizens of the country during the post-independence period a large number of universities and colleges were established without proper infrastructure and adequate resources. Moreover, with the galloping population growth there has occurred unprecedented increase in the number of students, leading to overcrowding in classrooms and hostels. In this kind of background a large number of persons were recruited as teachers in universities even though they did not possess adequate academic competence and were devoid of intrinsic motivation for acquisition and dissemination of knowledge and skill. With impoverished academic climate, malfunctioning administration and gradual corrosion of values and traditional roles, dissatisfaction among students increased and campuses became places of endemic strikes, agitations and indiscipline.

Secondly, in consonance with the spirit behind the national plans and schemes of development, there was a revolutionary shift in educational policy of the nation as a whole. It was resolved that education alone was not enough, it had to be socially relevant education, the traditional separation of the gown from the town was to be obliterated; and the students as well as the teachers were to be brought in the main stream of national growth as community development endeavours. Such a policy shift led to creation of National Cadet Corps and National Service Schemes. In order to implement the concept of a welfare state, a number of welfare schemes were initiated or broadened in scope for students, such as full-freeships or half-freeships, bursaries, scholarships, students-aid funds, health service programmes, etc. Industrial growth and technological advancement coupled with socio-economic changes in society led to rapid increase in the number and kinds of vocational opportunities available in the country. This happy development necessitated quick flow of enormous occupational information to the university students, so that they could plan and choose their careers. Governments established branches of employment bureaus on the campuses.

In this backdrop new tasks and functions emerged on campuses which were to be carried out along with teaching, research and academic management. To employ persons with expertise to handle these jobs would have proven to be more expensive than the governments were ready to incur. The bureaucrats looking after higher education and human resource development, not only found the most inexpensive solutions, but also instantly acceptable and apparently admirable justification for those solutions. The solutions were rather simple: assign these functions to some teachers to be carried out by them over and above their regular academic functions, and pay them extra remuneration for the work. Since all these new functions had emerged in the context of students, the tacitly assumed logic behind these solutions was that the teacher would carry the confidence and respect of students in performance of these tasks because of their moral and academic influence over them.

In order to maintain and promote discipline among students the proctorial board was enlarged and the Vice Chancellors were empowered to appoint with the approval of the executive council as many assistant proctors as they deemed necessary. Thus, in some universities there are more assistant proctors than the number of teaching departments. For the purpose of implementing student welfare schemes the dean of students welfare was given as many assistant deans as was considered necessary. Then, there is a chief of the employment bureau, and he is required to supervise circulation of occupational information and provide employment guidance. The national service scheme is run by a faculty who is appointed as its director and is assisted by a number of teachers working as programme organizers. For NCC a number of teachers are appointed officers after they are trained at some military training centres. Over and above these newly created extra-remunerative jobs, there are the traditional jobs of hostel wardens and superintendents, delegacy chairman and teacher in-charge getting fairly attractive remuneration and fringe-benefits such as free residential accommodation, telephone and orderly-cum-peons.

The statutory provisions for the aforesaid assignments have created a climate of extra-remuneration culture in the universities. A large number of teachers feel allured, and naturally enough, flock around the Vice Chancellors to get such jobs. Since Vice Chancellors have the power and predilection to treat the universities as their fiefdom and prefer to function as sheiks, they too create seasonal jobs for their favourite proteges. Such seasonal assignments are, such as, teachers-in-charge of re-evaluation, scrutiny and maintenance of examined answer books, of unfair means cases and corrections in tabulation charts, superintendents of examination, and co-ordinator for collection of marks and publication of results, directors and associates for admission tests to B.Ed., B.Sc., B.Com. and other classes. Over and above regular examination work and remuneration accruing thereof, there are recruitment tests sponsored by a number of organizations. Test administrators, supervisors and invigilators are appointed from amongst the university teachers also. As a matter of fact, the culture of extra-remunerative tasks has become so alluring and pervasive that a fairly large number of teachers remain busy in such jobs and have nothing worthwhile to do in the academic sphere in which they occupy their substantive positions. I know personally of a university which runs some diploma courses on 'no profit no loss' basis and charges all kinds of fees from students. Income thus accruing is not a part of the university budget and it is used to pay remuneration to the members of the teaching staff teaching the courses concerned, and to some functionaries of the administrative wing. In brief, the most conspicuous theme in university life now-a-days is the extra-remunerative assignment.

In creation of these extra-remunerative tasks, the most basic assumption, which has profound implications for the profession of teaching and research in Indian Universities, is that university teachers even after having completed the regular academic duties, have plenty of time left over for time consuming duties of proctors, deans of student welfare, directors of NSS, officers in NCC and Hostel Wardens and superintendents etc. In reality it so appears, but the way duties and responsibilities of university teachers have been conceptualized and the kind of complex intellectual functions these duties involve, how can a true teacher have four to five hours of time to be spared for extra-remunerative work in the university is nothing but a paradoxical puzzle.

Ruinous Impact of Extra-Remunerative Assignments

Ever since extra-remunerative assignments for university teachers have come in vogue, university work culture has completely collapsed and given way to a culture that fosters mediocrity and duty-dereliction, promotes extra-academic interests and ambitions, strengthens the hands of Vice Chancellors to exploit universities as their fiefdom, and makes the campus as the arena of rat-race for money at the cost of teaching and research.

Take any residential state university as a sample and the aforesaid picture will hit you in the eyes. You will find that appointment against all extra-remunerative jobs are made by Vice Chancellors. They, seldom pick up teachers who are committed to pursue the path of academic excellence. If such teachers are accidently picked up, most of them decline to accept because they consider these tasks more distractive and less rewarding. Only those teachers who have exhibited the potentiality or propensity for becoming ever-ready courtiers, have little intrinsic interest in reading, research and teaching, and excel in non-curricular operations and attainments, are eager to embrace these assignments. In practice thus appointed teachers do not carry out their academic duties sincerely and whole heartedly on the pretext of remaining overworked in their extra-remunerative duties. More often than not, such teachers become regular visitors of the vice-chancellor's court and keep him humored by concocting stories about the Vice Chancellor's gloriously effective method of running the university.

Mostly senior and middle-tier members of the faculties are assigned these extra-remunerative tasks. They remain more preoccupied with these tasks than with their academic work. Naturally enough, when the junior members of the faculties look up to their seniors for guidance and models, they find those senior colleagues, who earn more and work less, as more attractive models to emulate than those who work more but earn less and are seldom seen anywhere. Quite early in their career they discover that the only way to come in limelight and succeed in university life is to follow in the foot-prints of the exfra-remunerative job holding colleagues. Thus, for a fairly large number of teachers, academic duties become of secondary importance and coming in the goodbook of the Vice Chancellor and cultivation of courtier-like qualities acquire primary importance. New entrants as teachers soon discover that the adoption of the strategies of their senior colleagues, they would succeed in getting extrinsically rewarding assignments, and would be able to obtain promotions and elevation to higher ranks. On the other hand those teachers who are committed to academic work and values and keep away from manoeuvres and machinations for extrinsic gains, discover that within the university they become isolates and find their involvement in studies fairly frustrative. It is altogether a different matter that they find the work intrinsically satisfying leading to outside recognition, appreciation and sometimes fetching financial resources for further work. But it must be remembered that in terms of extrinsic gains in the university, such teachers remain poorer, neglected, and not infrequently punished also through denials of many facilities to them.

The assumptions, on the basis of which extra-remunerative functions and earmarked for teachers, have proven to be completely untenable and misconceived, ever since duties relating to maintenance of discipline and decorum, implementation of student welfare programmes, and promotion of national service schemes have been made extra-remunerative tasks for teachers to be carried out along with their primary duties of teaching and research, a pernicious kind of work ethos has made inroads in various state universities. Widespread cases of indiscipline, rowdyism, strikes, *sine die* closures of faculties and suspension of classes, and handing over of the campuses to police force have become endemic. Teachers have lost their moral influence over the student community. Ever since teachers have been assigned larger roles in monitoring examination processes and admission tests, academic sessions have been completely derailed and publication of results have become a long drawn affair. Despite a number of teachers working for students welfare activities, decisions on freeships, awards of scholarship, bursaries and all kinds of financial aids to students have become time consuming and unsatisfactory. In the case of national service scheme things have gone so bad that everything appears to be nothing more than mockery of national or community service. Camps are organized, sometimes on paper only, huge sums of money are spent, students get credit and certificates but what kind of national service spirit is inculcated in them is highly questionable. Barring rare cases of good work, most of the NSS activities are carried out in files and account books. It is a well known fact that ever since teachers have been allowed handsome allowances and fringe benefits for wardening and superintending the hostels, many hostels have become dens of undesirable elements. Large number of rooms are allotted on dubious considerations to persons who are more interested in political career and some in making money by donning the garb of studentship. It is not rare that some students continue to live in hostels and conduct all kinds of reprehensible activities and business from there. Wardens and superintendents enjoy the monetary and other benefits but ignore the hostels on the dubious plea of helplessness in the face of over-all deterioration of moral and administrative order in society. Whatever be the truth, it is undeniably true that the hostels remain in pathetic conditions.

Extra-Remunerative Tasks *versus* Academic Work

These two types of work in universities, *i.e.*, extra remunerative and academic, if not mutually exclusive, are at least poles apart, and require different kinds of motivations and skills. In all academic circles, especially in internationally reputed universities, it is believed that the members of academia have less time and more work to do, not because of any external pressures or compulsion, but because of intrinsic motivation and self-imposed commitment. If one does not have this kind motivation and self-discipline one cannot be a genuine academician. Sincere and serious academic work calls for total involvement with as little distructive atmosphere as can be created, and require freedom from daily travails and hassles of rat-race. It requires continuous nurturing and nourishing from the institution which employs teachers for academic fulfilment of material needs, rewards of appreciation and recognition of good academic work and congenial and peaceful work climate are essential ingredients for promoting academic pursuits.

These are, of course, ideal conditions and cannot be created in their entirety, but the guardians of universities should, at least, recognize these requirements and endeavour to fulfil them to the extent it can be done within the constraints of the overall economic and political conditions prevailing in the country. However, the author very strongly feel that by selectively providing extra-remunerative jobs to university teachers and that too in favour of academic mediocrity and non-academic accomplishments, is just the opposite of what ought to be done for promoting true academic work culture. Extra-remunerative jobs are administrative and managerial requiring verbal skills, physical ability and emotional maturity with integrity and impartiality. Academicians can do these jobs, provided that they have to be carried out as integral parts of day to day academic duties. Once these activities are separated from routine and made differentially rewarding and recognition and promotion fetching, they lead to creation of atmosphere in which intrinsically motivated academic tendencies are certain to be neglected and weakened. Such weakening is sure to be hastened if the work culture is underscored by pragmatism, expediency and arbitrariness. It is exactly the state of affairs that has on some university campuses these days.

If we consider what is expected of a teacher in the academic context. The University Grants Commission stipulates that each university teacher is expected to put in forty hours of work per week, of which twenty-four hours are to be spent in the department and sixteen hours of work for preparation and planning of teaching and research. Apart from institutional obligation, a university teacher has to keep on growing by acquainting himself with the newest knowledge that is coming in volumes in each discipline. A teacher cannot remain fresh, knowledgeable and effective unless he keeps himself abreast of the latest developments in his chosen field. If a teacher does all this, the natural question is: how much time does it leave for extra-remunerative jobs? No time at all. If one does these jobs, he has to neglect his academic duties and stop growing academically. Somehow-or-other a university teacher's job is such that it requires every moment he or she can spare, and still there will prevail a feeling of temporal insufficience.

At this stage of the analysis of extra-remunerative culture and its impact on the academic life in some state Universities, it should not be concluded by any stretch of imagination that university teachers are expected to be ethereal beings living in the world of ideas and ideals only and do not require money and material comfort, or they are expected to be completely immune to the influences of the world of contemporary realities. As a matter of fact, university teachers are as much of human beings as the members of any profession. They too have domestic responsibilities and social obligations to fulfil. They do need money to take care of their physiological and security needs. What is not expected of them is worship of the goddess of wealth. They are expected to be the intellectual cream of the society and paragons of morality, and they should be so perceived. What I have endeavoured to decry here is the corruptive inroads made by anti-academic extra-remunerative culture on the campuses and that too at the cost of academic values and ideals. The main thrust of this chapter is the extra-remunerative assignments to a few teachers for performing functions, which should be considered as an integral part of the educational processes and should have necessarily involved every teacher. What is condemnable about extra-remunerative assignments is their built-in potentiality for enhancing the reward and self-enhancement value and mediocrity and duty-dereliction culminating into pernicious impact on the core objectives of universities and emergence of a vicious atmosphere on the campuses.

It seems necessary to stress another point in the present context. University teaching is one of the noblest professions unique in every respect. This profession is like white clothes the whiteness of which must be zealously maintained otherwise dirt and dust of carelessness are sure to soil it. Universities are the organizations which mould precious human resources and university teachers are the human engineers. A sloppily constructed bridge is not likely to cause that big national calamity as poorly moulded human resources are sure to be. It is human moulding capacity and responsibility of teachers that is behind the policy of bringing the gown and the town together so that the gown could influence and provide guidance to the town. One of the most tragic thing that has happened on the campuses is that the gown itself has been influenced by the town and has commenced emulating the percepts and practices prevalent in the town. A large number of politics-oriented teachers now compare teaching profession unfavourably with a variety of professions and vehemently advocate for the same kind of work-culture and reward structure as is prevalent among some other professions. All these tragic trends have emerged on the campuses owing to unthoughtful and expedient changes in the higher educational system. These changes, especially introduction of extra-remunerative assignments for teachers, have played havoc with academic processes, ideals, and values and academic climate in a number of state universities has become extremely polluted and sickening. If something is not done immediately, the present trends will completely annihilate the very spirit of academic life in universities in a few years of time.

13

Roles and Responsibilities of University Teachers

In the last five decades Indian universities have been subjected to frequent systemic changes that have gradually augmented bureaucratic control over, and political interference in all facets of university functioning. Most of the structural and functional innovations promulgated by the bureaucrats under the guidance of politicians-in power have resulted in rapid erosion of academic standards and values, emergence of dysfunctional symptoms in the work-culture of campuses, and regular occurrences of highly undesirable activities all-around. In the context of the theme of this chapter, the two most insidious changes wrought upon universities are naively conceived fractionation of teachers' roles and responsibilities, and shift of emphasis from functions and functioning on forms and formalities in administration of academic affairs. The former has inevitably resulted in creation of extra-remunerative roles, such as those of proctors, examiners, deans of student welfare, in-charge and coordinators of different cells etc. The latter change has resulted in elaborate formalities in admission of students, completion of various examination processes, student union election and holding meetings of statutory authorities. All of these have been done at the cost of teaching and learning for which universities are established.

Under the present dispensations it is extremely easy to paint a portrait of the roles and responsibilities of university teachers. Once someone is appointed as a lecturer he/she has to teach those prescribed courses that are assigned to him, conduct research and publish papers, and somehow or other acquire externalized symbols of academic achievements such as awards, prizes etc. Apart from this he/she is required to do nothing else as a teacher. Of course, he/she may be asked to accept other assignments such as, be on the proctor boards, guest house care, students welfare office, superintendence of hostels and now Professor-in-charge of the mushroom like self-supporting programmes etc. for which there is extra remuneration. A teacher may or may not accept such assignments. Since all academic activities, be it learning, teaching or conducting supervising research, have become instrumental activities. A teacher has all opportunities to circumvent these activities by using effective shortcuts. These shortcuts in no way come in the path of a teacher's time bound promotion to higher posts.

The purpose of the present piece of writing is not to present a pen-portrait of what a university teacher does and is paid salary for. Here the cardinal purpose is to make an in-depth analysis of the concept of university teacher at an ideational plane, and to present an idealized account of the characteristics and distinguishing features of teachers categorized as lecturers, readers and professors, in the light of original assumptions on the basis of which university teachers were defined and delimited when Indian universities were first conceived.

Meaning and Categories of University Teacher

The term 'teacher' as a noun in English language is synonymous, with educator, tutor and master. Its Hindi equivalent is *'adhyapak'* now-a-days, and in ancient times it was *'Guru.'* In the context of universities its Hindi equivalent is *'pradhyapak.'* Each one of these terms implies specific characteristics. A teacher is assumed to possess an specific kind of roles to be enacted by him/her. As a generic term it refers to a person whose profession is to teach, instruct or to impart education.

A university teacher is appointed to acquire new knowledge and extend its boundaries and to disseminate knowledge and skill to undergraduate and post-graduate students. He/she delivers lectures in classrooms, supplements them by audio-visual aids, arranges for tutorials, conducts and supervises research, and keeps on enhancing his/her command over the medium of instruction and communication ability, along with his/her knowledge and ability.

Knowledge and skills are neither static nor ever complete. They continue to grow by virtue of incessant human efforts. As a matter of fact, in modern times, there is explosion of knowledge and skills in all disciplines. A university teacher is tacitly assumed to be committed to his/her calling and to continue studying newer contributions to his/her field of interest and motivation for continuous efforts at refreshing his/her repertoire of knowledge and skills.

University teachers fall in three categories. Generally a university teacher starts professional career as a lecturer. Then she/he endeavours to be appointed as a reader in her/his own university or elsewhere. The highest post in teaching profession is that of a professor. Here, some important questions may be raised. What are the distinguishing features of these cadres? Are these designations merely indicative of three pay scales to be given on the basis of the length of service put in as a university teacher? Are these cadres and pay scales created to dispel the possible sense of stagnation in career? Now that the so-called merit promotion or personal promotion scheme is operative in Indian universities, these cadres imply higher pay scales to be given on time-bound basis with a view to keeping university teachers immune from harmful feelings of stagnation. As a consequence of this scheme every lecture is assured (or insured, if you would) of rising to the rank of a reader after nine years of teaching experience, after this in eight years to the rank of Professor. Academic requirements for promotions have been completely watered down. In terms of academic merit the three cadres are not really distinguishable. One may claim to have twenty years of teaching experience, but it is possible that he has only one year's experience repeated twenty times over. Length of service is valuable but not as valuable as it is in administrative and business organization services.

The cadres of lecturers, readers and professors, as originally conceived, were basically required to do the same kind of work, but in other respects they were different. A reader is higher in rank than a lecturer and a reader is lower in rank than a professor. University teachers, irrespective of their designations are required to teach or deliver lectures in undergraduate as well as post-graduate classes, conduct and supervise research, and contribute their might in expanding the boundaries of knowledge and skill. A lecturer is primarily expected to be well versed in her/his discipline and in possession of requisite skills and expertise for effective communication with students in the classroom. Apart from his/her continuous efforts to keep abreast of newer developments in his/her area of specialization. A lecturer is also required to keep on sharpening her/his skills and augmenting her/his capabilities for supervising research.

Etymologically a reader means a printer's proof corrector employed to read manuscripts submitted for publication and to report on their quality and merit. In the context of universities a reader is required to be competent in critical appreciation of the known facts, theories, controversies and shortcomings in the area of his specialization in particular and in his discipline in general. Further more a reader is considered capable of new interpretation of the existing body of facts and theories, and draw fresh inferences encompassing new implications. It implies that a reader comprehends the subtleties of existing knowledge, gaps and contradictions inherent in them, and has the necessary competence and expertise to plan, conduct or supervise original research and publish it.

A highly coveted post for university teacher is that of a professor who professes his dedication and commitment to propagation and development of his/her discipline. A professor is expected not only to open up fresh areas of research but also to anticipate emergent trends in future developments. A professor is regarded as that breed of academicians who broaden the explanatory power of existing theories, and if necessary propound new theories and make innovations in research techniques. With scholarly insights a professor is expected to simplify apparently complex and abstract aspects of discipline concerned, and to pose new challenges for the future academicians.

The kind of characteristics and distinguishing features of university lecturers, readers and professors that are presented in the preceding paragraphs, were the criteria of filling in the posts of the three cadres a few decades back in India, and even today in the developed countries of the western world the same criteria are employed. These criteria may sound unrealistic and fanciful ideation in the context of Indian universities. The contemporary political ethos has led to emergence of such an educational socio-economic work-culture and value preference that the processes of learning, studying, teaching, research and students' performance evaluation are considered as merely instrumental activities to be performed as rituals for attaining the externalized goals of degrees, percentage of marks, Ph.D. theses completion and extra-remuneration. It is precisely the adaptation of the Indian educationists to this kind of ethos that these criteria are regarded as fanciful. Even when posts of Readers and Professors are filled by direct interviews, power politics, caste feelings, regionalism, and manipulative skills often prove more effective than genuine scholarship.

University Teachers: Models and Mentors

Students get enrolled and come to university campus with the objective of earning bachelor's or master's degree from different faculties and departments, and availing of opportunities provided by universities in extra-curricular fields, so that they may actualize their potentials in some of those activities. Students are overtly or covertly motivated to mature in academic environment so that they may grow up to become persons of such conduct, character and skills that are expected from highly educated individuals. All of these developmental characteristics emerge as outcomes of learning by students. They learn some of the characteristics intentionally. In any kind of social setup including those that characterize academic campuses, students intentionally acquire knowledge and skills by putting in their time and energy. However, they learn many more things incidentally. Such learning includes a variety of behaviours, as effective strategies for coping up with diverse facets of campus realities, academic norms, values, ideals, work-habits, and ethical requirements. It is indeed interesting to note that all human beings in general and students on campuses in particular learn a variety of behaviours incidentally by observing those who are considered superior, knowledgeable and worth emulation.

Fresh entrants to the portals of a university find its surroundings and work ethics unfamiliar. They do not know how to respond to unfamiliar settings and their requirements. They subliminally notice the diction, dress and deportment of their teachers and seniors and follow them as standards. Those teachers, who impress students by their scholarly lectures, effective communication and sympathetic treatment arouse spontaneous respect. Such teachers become models for their students, and are widely emulated. It is in this framework that every university teacher is required to conduct one-self in an exemplary manner. When behaviours of teachers reflect sincerity of purpose, devotion to duty, punctuality, and ethical concerns, students spontaneously develop respect for them, and through observation and imitation imbibe those qualities gradually in their habit repertoire. Teachers who are punctual and serious in class room, impartial as well as inspiring in their interactions with students and attentive and caring, command unstinted respect of students. University teachers are expected to function on the campus as friends, guides and philosophers. If teachers go to the university library for regular study, students also develop habits of going to the library and studying. Thus, university teachers are professionally required to function as models for students.

One of the roles, and for that matter the most important role, of a teacher is to be a mentor for students under their charge. Engineers build houses, roads and bridges, physicians cure diseases and prescribe for healthy life, while university teachers mould and shape young human beings and transform them in precious human resource for all round development of society. If a bridge is carelessly built and collapses, a few crores of rupees and a number of individual lives are lost, but if teachers become careless and are given to surreptitious evasions of their roles and responsibilities, numerous generations of human resources are malmanufactured, and the entire country gets disturbed and dysfunctional.

Effective university education hinges on the quality of mentoring programmes a university has included in its academic calendar. In this context mentoring means that a teacher takes twenty to thirty students under his/her wings, meets them once a week in groups or individually, finds out their problems that interfere in their curricular and career activities, and then help them in solving those problems. These problems may be directly related to their curricular activities, such as poor comprehension, faulty method of study, anxiety caused by discrepancy between their levels of achievements and aspirations, non-availability of relevant study materials, lack of career consciousness and planning, indigent background and inadequate financial resources. It is a responsibility of the mentor to help their protéges in solving these problems, not necessarily by spoon feeding but by creating insights in them to find solutions to their problems.

The role of a mentor, and responsibilities emanating from it, cannot be performed effectively unless the mentor has certain personality characteristics as a consequence of his/her natural development as a human being or of his/her deliberate endeavours to imbibe certain values and behavioural habits. A mentor is one who possesses the power of positive thinking, high self-esteem or a sense of self-worth, and is competent as well as effective. A mentor is primarily an other-oriented individual who gives and helps those who come in his/her contact professionally. In order to become an effective architect of human resource, a teacher must have opted for teaching profession as matter of free choice because of his/her unshakable belief that knowledge and expertise are power more potent than any other source of power.

Teaching profession, especially in institutions of higher learning, is like white set of dresses that symbolizes purity and erudition. The goddess of learning, Saraswati is always dressed in white fabrics. It is obvious that one dressed in white clothes has to be extra-careful for maintaining its whiteness because it is easily vulnerable to get sullied and tarnished, especially during modern times of polluting and corrupting forces. Mud slinging and character assassination of fellow teachers that is done out of professional jealousy must be stopped. Such activities are fatal to the educational environment. Naturally enough, a university teacher has to be extra-careful in his/her conduct, character and extra-professional activities. When a university teacher is out on the campus going from one building to the other or in the surrounding community, he/she gets noticed and spectators talk about his/her qualities of head and heart. So every university teacher has to be extra-careful and must not do anything, such as apathy for students, lack of interests in students outside the classrooms, intolerance and negativistic interest, avoidance of involvement in maintenance of discipline, irregularity in performance of her/his duties and favouritism etc. A university teacher is expected to have simple lifestyle and high thinking. Such a teacher can be a model and mentor of students.

Skills

It is a known fact that a university teacher's work is quite complex, demanding high level of intellectual capacity, modelling—mentoring skills and a variety of expertise. A teacher has to be a knowledgeable scholar of his discipline and must have numerous skills. It goes without saying that a teacher must have a good command over the language, which happens to be the medium of instruction. However, the most important skill a teacher must possess is that of effective communication. Teaching involves more or less two-way communication between teacher and students. A teacher is required to transmit information orally to students at a level that must be easily grasped by all students, and has to be interesting enough to sustain their attention. For rich encoding and storage of transmitted information by students a teacher must be skilled enough to use audio-visual aids for supplementing oral communications. Built in redundancies in oral communication with statements of critical points, and multiple examples of abstractions go a long way in making teaching effective and attention sustaining.

Every discipline of knowledge focuses on some specified aspect of the universe in which human beings live. Each one of them acquires knowledge and understanding about those aspects that have vital implications for simplifying and ameliorating the conditions of human survival and comfort. It is in this context that application skill becomes relevant. This skill refers to human ability to discern between facts, phenomena and principles delineated in a discipline on one hand and objects, events and confusing ambiguities that occur in real life.

14

Indian Universities: Underlying Assumptions and Consequences

The Indian universities, whether residential or affiliated, Central or State and unitary or federal, are, at present, in an unprecedented state of turmoil and turbulence giving rise to precipitous academic decadence. The long cherished values of devotion to duty, commitment to the pursuit of academic excellence in acquisition, dissemination and extension of knowledge and skill, intrinsic motivation for being competitive in one's chosen field of study and unconditional respect for merit have been gradually thrown in the limbo of oblivion. Corruption in financial management, unethical practices in examination processes, shortsighted expediency considerations in day-to-day administrative decisions, surreptitious use of short-cuts in teaching and research, undue emphasis on examination with little concern for effective teaching, politicisation of socio-academic processes, disappearance of co-curricular and extra-curricular activities and interference by bureaucrats and politicians in power have currently become the hallmark of university education. Work is no more alluring than the benefits of knowledge and skill development.

The basic question to be asked is: What factors have predisposed university functioning to this state of deterioration? In order to find out the factors behind the malfunctioning of university academia one has to analyse the structural changes that have been brought about during the last five decades and the assumptions that underlie those changes. It is an undeniable fact that the present day structural and functional features of universities have largely been shaped by the socio-political ideologies and compulsions in the country. The national obsession with economic and industrial growth and disappearance of morality from the socio-political ethos have not spared the campuses from their deleterious impacts. However, the most pernicious causative factors are the wide ranging shifts in the basic assumptions on the basis of which universities have been structured. It will be a worthwhile exercise to examine those assumptions and then scrutinize their effects on the universities in this country.

Historical Overview

The present system of education from primary schools to universities originated and developed when the entire Indian subcontinent was under the British rule. It was originally conceived and developed by Lord Macaulay for the citizens of a subjugated country. They were to be trained and employed as minor functionaries in governance of the country, but were not to be trusted. That is why the administrative set-up in universities was hierarchically organised. For no administrative functions, such as general administration, accounts, examinations, engineering works, etc., a single individual was empowered to take the final decision. Every matter had to be initiated at the bottom rung of the administrative ladder and to be moved to the top rung for final disposal. The top rung consisted of Englishmen, Anglo-Indians or Indians of proven loyalty to the crown. No single subordinate was considered trustworthy enough to be empowered for final disposal of any matter. That kind of administrative set up is still in operation with further tightening of bureaucratic control. However, the academic wings of universities were shaped according to structural and functional patterns prevalent in the universities of England and Scotland.

During the pre-independence era, the major objective of establishing universities by the Central or State governments, philanthropic organisations or social service-oriented individuals, such as Mahamana Madan Mohan Malaviya, Sir Sayed Ahmed, Doctor Hari Singh Gaur, etc., was to provide space, buildings and necessary resources for acquisition and dissemination of knowledge and skills with the tacit assumption that to become knowledgeable and skilled was to become good and efficient. However, in the 1950s, two significant events occurred which drastically changed the concept of university education in India. First, after the end of Second World War, a distinction was made between knowledge and utilitarian knowledge and also between education and relevant education. The second event that had direct influence on education was the emergence of the sovereign state of India, freed from the British rule.

As per the provisions of the Constitution of India, higher education became a concurrent subject of the Central and State governments. Thus, the responsibility of establishing, providing administrative structure and financing universities became a function of the governments. In accordance with the national goals of economic-industrial development and social change, separate institutions of relevant and utilitarian education such as IIMs, IITs and national laboratories were established on the pattern of developed countries. This step led to increased emphasis on job-oriented professional courses that would ensure subsequent employment to the outgoing students. Naturally, enough universities in this country were called upon to evolve such courses on studies that would be subsequently utilized by degree and diploma holders to function as a human resource for socio-economic development.

Since establishing, maintaining and controlling university education became the government's business, the University Grants Commission (UGC) was created for funding universities and, later on, also for monitoring academic standards as well as for implementation of directives of the Central Government. Since then, frequent structural and functional innovations have been made by the successive governments in the light of socio-political ideologies of politicians in power. As a result, university functions have changed. More often than not these changes have led to the present state of affairs in Indian universities. Let us examine the major changes introduced so far.

Interaction between Gown and Town

Owing to major emphasis on relevant and utilitarian

education, the nature of functional relationship between university dons and society at large has been redefined. Since the very inception of the modern universities, university dons used to effortfully maintain a distance from the surrounding communities and they were characterized as living in ivory towers far removed from the humdrum of public life. The idea implicit in this arrangement was that the dons with their ideas, ideals and published work would influence society at large indirectly, but would remain unaffected by it. It was assumed that the students after passing out of universities would function as model citizens in society, to be emulated by others.

With onward march of the popularity of relevant education as an instrument of human resource development and social change, institutionalised attempts at promoting intimate interaction between the gown and town were introduced. Such attempts are best illustrated by establishment of a variety of activity centres, such as the departments of National Social Service Scheme, Adult Education, etc. At present, extension of knowledge and skills has become an important aspect of university functioning. The Central as well as the State governments regularly exhort the academia to study the problems of surrounding communities and function as instruments of social change.

The most pertinent question is: How has the newly introduced system of intimate interaction between the academia and the communities affected university functioning? Have the university dons succeeded in spreading literacy in the masses? Have they been able to create a sense of social responsibility and make the masses aware of different kinds of social traps they have been unknowingly forming and, thereby augmenting social dilemmas? Is it not a fact that the academic community itself has become victim of the ideas, ideals and ways of society at large instead of influencing and modifying it? Is it also not true that the present day university communities have got involved in the rat race of money making, pursuits of power and promotion just like the commoners, instead of remaining confined to academic pursuits? These questions need to be thoroughly examined and the relationship between the gown and town needs to be redefined.

University Personnel as Non-Trustworthy as the Commoners

If the structural and procedural amendments in the University Acts and Statutes introduced by the successive governments during the past five decades are carefully scrutinized, it becomes clear that university autonomy, which used to be the mainstay of higher education at a time when the country was under foreign rule, has been withdrawn. Is it not paradoxical that when there was no freedom in the country, university senates and/or syndicate (often called courts and executive councils) were empowered to get persons of their choice appointed as Vice-Chancellors? All academic and administrative personnel were appointed by internally formed committees with no interference from the governments. Universities enjoyed complete autonomy in managing their financial affairs. But soon after the country became an independent sovereign state, university autonomy was gradually liquidated and university functioning came under direct control of the Central and the State governments. Teaching departments were entrusted the responsibility of developing their respective disciplines of knowledge and skills through research, but the teachers were not considered trustworthy enough to select new colleagues. Formal selection committees had been provided for, in which even the departmental heads have no decisive role to play and more often than not, Vice-Chancellors in collusion with external members have the final word and appointments are made on various extraneous considerations because no one now feels accountable for poor selections.

Once upon a time, all teaching members of various departments were entrusted with the duty to evolve syllabus outlines and the course content of each paper was prepared prior to the commencement of the academic session by the teacher who had to teach that paper. At present, this kind of autonomy cannot even be dreamed of. Now-a-days, formally constituted boards of studies write syllabi. In some states, uniform syllabi are prepared for undergraduate classes and all universities in that state are bound to teach them.

The administrative set up of the universities is basically the same as designed by Lord Macaulay but has been further tightened. Its net result is that with an increasing number of students and enhanced complexity of academic requirements as well as formalities, university administration has become cumbersome, time consuming and non-responsive to academic requirements. Almost all decisions are taken by Vice-Chancellors who are appointed on grounds of their political affiliations and manoeuvrings, instead of their academic credentials, dedication to the cause of education and vision. Now, Vice-Chancellors are in no way accountable to the academia but, only superficially to the executive councils which are packed with non-academicians with political clouts along with teachers by rotation. Thus university administration is accountable to none for its arbitrary ways. At present administrative responsibility is so diffused that no one feels guilty about the chaotic state of university management. The total number of administrative staff is twice more than the number of teachers in every university but administration has become completely inefficient and irresponsible.

In all universities, barring some exceptions, there are boards of studies, one for each discipline. These boards consist of some internal members and two to three external experts. The meetings are convened and chaired by heads of the departments concerned. These boards prepare panels of paper setters and examiners, of which approximately half of the names have to be from other universities. The Vice-Chancellors randomly or arbitrarily tick required number of names. Paper-setting, evaluation of answer scripts and preparation of results are not only totally divorced from the teaching-learning processes, but have also become secret and confidential affairs, to be managed with utmost anonymity. Such a procedure has made academics examination-heavy and teaching-learning light and both these activities are now conducted as rituals. Widespread use of unfair means in examination halls, over or under-evaluations of answer scripts by examiners on extraneous and reprehensible considerations, tampering with entries in tabulation charts and issuing fake mark-sheets have now become matters of common practice and public knowledge.

The implicitly assumed non-trustworthiness of university personnel is very vividly mirrored in the bureaucratically prescribed procedure of preparation and finalization of annual budgets. Instead of giving block grants, the governments require universities to prepare annual budgets in detail and then, the bureaucrats in the government prune, modify and rearrange them. More often than not, even the most genuine and urgent demands are expunged. Thus, instead of having financial autonomy in managing their financial affairs, universities are sanctioned lopsided and unrealistically reframed budgets which more often than not retard genuine and useful academic developments and sometimes pamper spurious university plans and projects.

Objective Criteria for Assessment of Academic Merit

According to the Constitution of India, one of the fundamental duties of Indian citizens is to develop scientific temper. It is implied that every Indian would develop an appropriate mental set to perceive and understand other persons, events occurring around him and all other objects in an objective manner so that one's understanding and assessments are not coloured and distorted by one's prejudices and preferences. In the realm of higher education, the principle of objective criteria for judging an individual's academic merit has been mechanically applied without ever having pondered over its disastrous consequences.

Unlike the universities in developed countries and unlike our own practice during the pre-independence era, now merit is judged on the basis of one's certificates and degrees, percentage of marks in various examinations, number of published research papers, monographs/books, etc. and length of teaching experience. Thus, academic merit and attainments have now been transformed into externalized symbols. How those symbols have been acquired and from what kind of sources, are considered irrelevant. The essential qualification for becoming eligible for lecturership in a degree college or university is successful clearance of National Education Test (NET) and consistently good academic record with more than 55 per cent marks in Master's degree examination. This illustrates the externalisation process. Naturally enough, the new goal set for students is to get high percentage of marks, acquire degrees and to have as many published papers to their credit as they can manage without bothering about their academic worth and quality.

The assessment of academic merit and accomplishment in terms of externalized symbols has made educational processes an instrumental activity. It is a fact that in whatsoever realm of public life and at whatsoever time, if and when goals to be attained are externalized, either in symbolic or concrete forms and if socio-political ethos is morality neutral, virtue evasive and devoid of social surveillance, innate predispositions of human beings force them to use short cuts in order to attain those goals. In this backdrop it is not surprising that ever since measures of scholastic accomplishments have been objectified and externalized in symbolic forms, students, teachers and educational administrators have become strongly prone to use short cuts in studies, teaching, completing courses of study, answering questions in examination halls, evaluation of scripts, supervision and award of doctoral degrees and also in publication of research papers. Today, ornamental symbols of scholarship have become success fetching measures. One may be a scholar of extraordinary attainments, but he/she is not legally qualified for even lecturership unless and until he/she has externalized symbols of merit in possession.

University Education: An Instrument of Social Justice and Equal Opportunity to All

Educational structure throughout the world is conceptualized as pyramid shaped in which university education constitutes the top. It is made available to only those students who are endowed with higher mental abilities, proper scholastic aptitude and superior motivation for academic activities. The underlying assumption is that the processes of higher education are too complex as well as too abstract to be successfully negotiated by all higher secondary school certificate holders. That is why even in the highly developed nations of the world a very small percentage of students are accepted as regular students in universities. Nowhere in the advanced countries is higher education treated as an instrument of social justice. Academic criteria of admissions are not lowered or brushed aside for the sake of social justice or equal opportunity to all.

However, in India, an entirely variant concept of university education is being practised. Soon after attainment of independence, India opted for a democratic form of government and socialistic pattern of society in which all citizens, irrespective of caste, creed, locality and social status were assured equality of opportunity in various spheres of national and social life. From the vantage point of constitutional provisions, Indian society was discovered to be non-congenial for emergence of a just social order. In order to remove certain oddities and built-in-infirmities, the Indian polity had to go through a series of turmoils and tensions. Ultimately, politicians in power and bureaucrats as their loyal cohorts found a solution in the concept of reservation of seats and posts for the perennially downtrodden scheduled castes and scheduled tribes in the institutions of higher learning. Later on, this facility was extended to other backward castes. At present, approximately fifty per cent seats as well as administrative and academic posts are reserved for candidates of the aforesaid categories. There is no denying the fact that now the monster of social injustice has been killed but, simultaneously, this step has made academic excellence as an anathema to social justice and equal opportunities. This non-imaginative step has also led to creation of another national trap in which academic standards and healthy competition have been debilitated.

With the onset of planned activities for all-round national development, demand for skilled workers, technocrats, business executives and civil administrators increased by leaps and bounds. The public demand for opportunities of higher education to their wards motivated Central and State governments to establish new universities and encouraged philanthropic organisations and individuals to establish new colleges. As a result, in the first three decades, the number of universities increased from 20 to more than 200, most of which were having large number of affiliated degree and post-graduate colleges. Such a condition gave rise to two highly ominous phenomena in the world of higher education which were now embedded in a morality-neutral, power-seeking and money-oriented socio-political ethos. First, an enormous number of undeserving persons got easy entry in the portals of universities and colleges. Second, the campuses became overcrowded because the authorities had to succumb to different kinds of pressures—often coercive and sometimes alluring. Seats in various classes were increased beyond institutional capacities.

In this backdrop, a subtle shift in higher educational policy has occurred. Since the last few years, the State governments are encouraging establishment of self-financed colleges even in remote rural areas for the local students. They are allowed affiliations to universities even though they have no proper buildings, libraries, laboratories and sports grounds. Such colleges are free to appoint lecturers on terms and conditions set by the respective managements. They enrol anyone who holds a higher secondary school certificate. Politicians in power, bureaucrats in control and public at large are contented that the noble cause of equal opportunities of higher education to all is being satisfactorily served. However, the overall outcome of such policies is that we have started manufacturing unemployable and half baked graduates who are fast becoming a source of a variety of social problems, which are gradually augmenting the woes of communities.

University Service Modelled as Government Service

Owing to the fast changing political and economic scenario, university service conditions have been modelled after government run departments on demands of university teachers' associations. It is well known that from the early days of independence, leaders of teaching community started functioning as trade unions, adopting agitational approach to secure parity with service conditions prevalent in government services at the cost of university autonomy. Its net result is that now university service is directly controlled by the Central/State governments and university teachers have been given hefty pay scales, merit (really time bound) promotions and impregnable security in service conditions. Now, universities have, more or less, acquired the status of government-run departments.

These changes had disastrous effects on academic processes. For example, consider the case of a university lecturer. As soon as one is appointed as a lecturer, he/she is assured (or insured !) of being elevated to the status of a professor even though he/she does not put in a single day's honest and sincere work, because on grounds of merit in theory and of the length of service in practice, promotion is an important service condition. Naturally enough, if one can be elevated to the position of a professor on a time bound basis, why should one be motivated to burn midnight oil and indulge in the arduous task of acquiring up-to-date knowledge and extending its boundaries through research. The provision of time bound promotion has drastically reduced the chances of university teachers moving from one university to the other and has contributed to in-breeding. Moreover, this provision has undermined teachers'motivation to be competent and recognition seeking and to be competitive as well as preservative.

Fractionation of a Teacher's Role and Responsibility

The gradually increasing number of structural and functional innovations introduced by government over the years have made a teacher's role and responsibility highly demanding and complicated. Now, instead of requiring teachers to perform all those functions, the roles and responsibilities have been subjected to fractionation. Ordinarily, university teachers are required to teach, conduct as well as supervise research, provide correct guidance to students, act as models as well as mentors and motivate them to develop a sense of social service and civic responsibility and finally, are required to evaluate students' scholastic and extra academic achievements on the basis of which diplomas and degrees are to be awarded. But, at present, many of the aforesaid roles have been taken away and assigned to a set of new teams created for this purpose, such as the Vocational Guidance Centres, National Social Service or to specially created teams of teachers on payment of extra remuneration. Such teams include Proctorial Boards, Dean of Students' Welfare with numerous Assistant Deans, Centre Superintendents with numerous invigilators and quite often, 'Flying Squads' for safeguarding sanctity and fairness of examinations.

The consequences of such fractionation have further vitiated and distorted academic life on campuses. The Proctorial Board usually consists of one Chief Proctor and numerous Assistant Proctors representing different faculties and each member of it gets handsome extra remuneration for maintaining discipline on the campus. However, students' discipline has become complicated and the members of the Proctorial Board have to remain busy and vigilant during working hours. This they do at the cost of teaching. Such an arrangement inevitably makes other teachers rightly assume that they are not supposed to be involved in discipline maintenance. In the same manner, common teachers assume that students' welfare, career guidance and social responsibility development are areas of specially created teams. Naturally enough, teachers who are not paid extra remuneration for maintenance of discipline, career guidance, students welfare, etc. tacitly assume that they are institutionally not required to get involved in such activities and remain confined to teaching if and when they are required to.

The most harmful change is the separation of the students' achievement evaluation from the learning and teaching functions and undue emphasis on the importance of marks obtained in examinations. Treating the examination process as separate from teaching and making it an extra remunerative function is typically an Indian practice and in no developed country is such a separation in vogue.

Universities as Nurseries of Future Politicians

Originally one of the functions of universities in India was to provide opportunities, facilities and organisational structure so that students could develop skills necessary for efficient group activities of co-curricular and extra-curricular types and gain experience in democratic ways of doing things of common interest. Much before this country was finally liberated from the British rule, Mahatma Gandhi exhorted students to come out of their academic campuses and add their strength to accelerate the pace of freedom struggle. Though this call was for an extremely noble cause, it added political flavour and fervour to the activities of students' unions. After independence and with increasing control of universities by the government, all political parties of the countries created youth wings of their parties and universities provided fertile grounds for expansion and harvest of human resource for political rallies, broadening public base and election campaigns. In mid-seventies the political agitation launched by Jaiprakash Narayan for total revolution in the country transformed university campuses as miniature theatres of politics. Since then, the students' union elections have become testing grounds for political leaders in which they assess their strength.

As a consequence of the patronage of politicians and political parties and due to the large scale criminalisation of politics, students, union office bearers, as well as defeated student leaders use university life for getting well trained in the techniques and tricks of the politicians. Instead of contributing constructively to co-curricular and extracurricular activities, students' leaders function as trade union leaders and fight for increase of seats for admission, extraction of money from university for furnishing and maintenance of union buildings, pressurizing authorities and teachers for undue favours to their followers and more often than not function as touts for getting fake mark-sheets, degrees and other advantages from university administration. A large number of socially undesirable characters take admissions on fake credentials in spurious kinds of diploma courses or Ph.D. programmes, get accommodation in hostels either in collusion with, or with the connivance of admission committees and hostel authorities. Such students, instead of pursuing studies for academic advancement, do everything necessary to get framed for successful political careers or to become mafia dons. It is now self-evident the more the tightening of bureaucratic control over the university, the more the student community becomes prone to use pressure tactics, unethical practices and dubious methods of working for their political growth.

In conclusion it may be noted that the current state of affairs in most of the Indian universities is frightening as well as depressing. Most of the assumptions on the basis of which universities have been restructured are not necessarily unjustifiable, but the way they have been implemented has proven to be deleterious and has led to the corrosion of the basic nature and values of higher education. No sensible person can deny the importance and value of social justice and equal opportunity to meritorious students, but how to create a socially just system in this country needs serious thinking with a view to developing it without compromising on the basic values of university education. Academic excellence and social justice are basically not opposed to each other.

All intellectual pursuits thrive only in genuinely democratic and open systems. In order to develop universities and colleges that encourage and support scholastic achievements, intrinsic motivation for intellectual pursuits and development of socially useful expertise and academic excellence, the present day universities will have to be overhauled. The most urgent need of the hour is to initiate a national debate for redesigning universities that may create valuable human resource and promote the onward march of knowledge expansion and skill development. Universities must cease to be the manufacturers of unemployable degree and diploma holders and must be freed from political and bureaucratic control. Higher education must not be as cheap as it is today, but students with indifferent economic background should be provided with more liberal support. Total dependence of universities on Central/State governments has to be abolished and at the same time, full autonomy of universities must be restored. If politicians, whether in power or in opposition, can be trusted to govern, academicians and social service oriented intellectuals and industrialists—philanthropists can be considered conscientious and trustworthy enough to manage universities.

15

My Concept of a University

After having served as a teacher for four decades in a residential-cum-affiliating university, and also after having played quite intimate participative roles in academic activities of a large number of State as well as Central universities, I have no hesitation in asserting that Indian universities have now been derailed and fallen in abysmal depth of malfunctioning. Deterioration is so alarming that all intellectuals, social workers, academicians, and philanthropists should get up and start a national debate for idealizing universities that function as centres of excellence in acquisition, dissemination and extension of knowledge and skill development, with a view to producing not only scholars, technocrats and doctors, but also conscientious, industrious and competitive citizens for creating a truly secular-democratic India.

Functional Domain of Universities

At present universities in India do not have a uniform character and variance in their forms and functions is so wide-ranging that no two universities are comparable. Keeping in view enormous population and prevailing zeitgeist in India, clear-cut distinction must be made between arrangements for undergraduate and post-graduate education. Education up to the bachelor's degree in humanities, social, physical, life and actual sciences must be considered as the highest phase of general education and be made available to as many higher secondary school certificate holders as are keen to study as regular students. It should be taken care of by degree colleges under the umbrella of some affiliating universities. Residential universities must not be burdened with affiliated colleges in the same way, as the affiliating universities must be allowed to venture into starting residential wings in their campuses. There may be autonomous colleges having both undergraduate and graduate courses of study in basic disciplines. However, bachelor's degree is and should be considered as the optimum level of education so far as the various types of ministerial and administrative jobs are concerned. Post-graduate degrees are essential for teaching and other academic jobs.

In defining the functional domain of university, it is necessary to maintain distinctions between basic and applied disciplines. Basic disciplines are end in themselves and must be linked to intrinsically motivated activities that have to be pleasurable in themselves, while applied disciplines are means for shaping competent human resources so that all facets of human life may be made more comfortable and pleasurable. On the other hand, in basic disciplines it takes decades to make worthwhile break-through in extending the boundary of knowledge, while in applied disciplines annual evaluation becomes necessary for assessing the effectiveness of the courses in terms of the quality and quality of employment outgoing students are able to get, and even newer developments are quicker. Because of these vital differences between these two sets of disciplines, I very strongly feel that no institution or university should be burdened with the task of developing both.

Forms of Universities

My concept of university is that it must have a single identifying characteristic and must assume one of the many available forms, such as, residential, affiliating, federal and open. To have hodge-podge of these four forms is ruinous to academic excellence-building. Applied sciences, and technological disciplines classified under the subject of relevant education, best thrive in separate institutes as is evident from IITs, IIMs, AIIMS etc. To me a traditional university must be residential, whether unitary or federal, and should run on the basis of twenty-four hours of curricular, co-curricular and extracurricular programmes. Students should have residential facilities in hostels and hired lodges or if allowed to live in their own residences or hired rooms, should have delegacy facilities so that studentship becomes a twenty-four hour affair under supervision of university teachers. An internationally renowned professor once remarked that a good university makes students learn more in the corridors of campuses than the classroom.

An affiliating university should function as a coordinating centre for colleges, undergraduate and post-graduate classes for academic purposes and administrative uniformity, and help in establishment of colleges in smaller cities and towns without compromising on academic as well as extra-academic activities and simultaneously make it possible for meritorious students from remote areas of the country to get university education. Only open universities should be allowed to run correspondence courses and provide for distant education.

The main objective of establishing some university is to create a setting, a climate and a culture of genuine acquisition, dissemination and extension of knowledge as well as skills so that the country may become rich in skilled human resource, law-abiding, industrious and creative citizens as well as peace-loving human beings. Cumulative historical experience of higher education all over the world suggests that academic excellence sprouts up and flourishes in systems that are interactive and open, intrinsically motivating and merit-appreciative but face scholarship depreciative. Such a system must be free from favouritism, expediency-considerations and discriminations on the grounds of caste, creed, sex, socio-economic status and locality. With a view to creating positive climate and genuinely productive work-culture in universities must be structured and function on the basis of certain assumptions.

Trustworthiness of University Personnel

At present the academic as well as the administrative wings of universities are structure on the assumption that every one serving in the university, whether as a teacher, administrative officer or office-assistant does not matter, must not be trusted. Such an assumption is legacy from the times India was under

British rule. This assumption has been further entrenched during the last five decades by the morality-neutral politicians in power and red-tapism-loving bureaucrats. It is precisely under this assumption that the university administrative set up is vertically organized. That is why every paper whether from the academic sides or from public, has to pass through a number of tables before the final decision is made. Like-wise for a variety of academic matters no trust is reposed in teachers. It is one of the reasons that examination has been divorced from teaching, and elaborate rules have been framed for appointing external examiners, creation of secret and confidential sections for examinations. Such a mode of functioning has made university administration cumbersome, inefficient, time-consuming, and bereft of accountability and an area of diffused responsibility. Academic functioning too has become examination-heavy and teaching-light.

In all democratically developed countries all members of the university personnel are assumed to be trustworthy and everyone's domain of duties and responsibilities are clearly defined and delimited. Each person, whether an office assistant or administrative officer and whether a lecturer, reader or professor, is empowered to make decisions concerning problems and issues arising in the domain assigned to him/her and is accountable for the same to the section-in-charge, dean concerned or chairman of the departmental council.

The top echelon of a university, *i.e.*, Vice-Chancellors, Deans, Professors, Finance Officers, Registrar etc. should have responsibility of creating healthy work climate and culture by their exemplary style of functioning. Trust and accountability reposed in university personnel provide germane ground for the growth of self-esteem and a sense of belongingness. These qualities pave the ground for sincerity, efficiency and commitment to one's duties provided that the top officials and academic dons are punctual, regular, just and efficient, as well as maintain the dignity of high offices they hold, without ever succumbing to unjust pressures and expediency considerations. It is a universally acknowledged fact that favouritism, corruption, slackness in duty, misuses of power and degeneration of values originate at the top and filter down to the bottom of the employees. Likewise impartiality, honesty, devotion to duty, judicious use of power and virtuous values originate at the top and permeate the university campuses.

A Vice-Chancellor as a competent administrator and also as an educationist of vision must be trusted and empowered to develop his/her university by showing the qualities of a hard task master in organizing academic work, setting standards and focusing on goal attainments. He/she is rightly expected to be an effective social leader too at building team-work, mediating conflicts and providing impartial support to all those who serve in or are concerned with university functioning. A vice-chancellor should be fully accountable to the executive council, the apex authority in the university. In the same way the Deans, Heads of Departments, Registrar and alike are trusted and empowered to function autonomously in their domains of duties and responsibilities.

Trust as the basis of distribution of work areas and duties makes work economical in terms of required manpower and saves time. It also promotes efficiency and competitiveness. Trust paired with accountability lays the foundations for creating genuine work culture and moving forward to healthy and responsible administrative set-up and academic excellence. This point can be usefully illustrated by a recent event that occurred in one of the State universities. In students union election the candidate who got elected as president, was discovered to be a non-student during that academic session. This discovery made by one of the defeated candidates led to large-scale agitation by student leaders opposed to that person. As a consequence the entire administrative work got paralyzed for a few days. In the mean time official probe in the matter revealed that the elected candidate was really not a regular *bonafide* student. Is it not ironical that the members of the election committee scrutinized his nomination papers and he had been found to be a *bonafide* student. However, instead of holding the members of the elections committee accountable for this blatant error, his election result was cancelled, and next highest votes getter was declared to be elected. This incident provides a glaring example of responsibility diffusion for acts of omissions and commissions in the university set up because no one is considered accountable for one's actions and inactions.

University Autonomy and Freedom from External Control

Universities that are internationally recognized and eulogized for their contributions in expanding the boundaries of knowledge, scientific discoveries and technological innovations, are fully autonomous and absolutely free from bureaucratic control and political interference. They are financially as well as administratively independent. Some of these universities do accept financial grants from federal (central) and/or state governments but without strings attached. These universities charge handsome amounts as tuition and other fees because higher education is considered an essential commodity for public distribution.

Ever since universities in India have come under direct control of State and Central governments, they have been reduced to the status of government departments. Stringent formalization of procedures, abject dependence of universities on bureaucrats, and direct or indirect but frequent interference by morality-neutral and populist-politicians have made university life utterly form-heavy and content impoverished sidetracking the universally cherished motto of academic excellence.

An autonomous university constitutes its court (senate) which may consist of some elected representatives of its faculties, rich donors, a few eminent scholars, scientists from other universities or institutes, distinguished alumni, educational administrators and technocrats. A search committee constituted of five court members of the court would propose five names from which the executive council may pick up one to be appointed Vice-Chancellor by the Chancellor. Thus, appointed Vice-Chancellors may be given five years to function in the first instance. Extensions may be given for indefinite periods on the basis of recommendations to be made by the university court. Other administrative officers are to be appointed by the executive on a tenure basis.

Appointment of Professors, Associate Professors, Readers and Lecturers through the current mode of selection must be redesignded. Whenever a university is in need of recruiting fresh academic staff, the University Registrar, in consultation with the Chairperson of the department concerned, send to the Registrars of all universities a few copies of the advertisement for the relevant departments there, so that aspirants for those posts may apply with detailed curricular vitae. The posts may also be advertised in professional journals of the disciplines concerned. Applications thus received should be scrutinized by the Professors in the department and short-listed. References about short-listed candidates should be obtained and two or three prospective appointees may be invited to give seminars for a week in the department to be audited by senior staff members. After such processing of the academic credentials a person may be given appointment on a salary commensurate with his scholastic achievement for a short period and on successful completion of this period, he/she may be given tenure appointment.

It is very pertinent to point out that the academic merit of a candidate must not be judged on the basis of objective and externalized symbols of merit such as percentage of marks, degrees, number of published papers and teaching experience etc. These appendages are neither reliable nor valid indicators of one's merit. Sincere, committed and time-tested academicians only can truly judge anyone's merit on the basis of prolonged interaction in academic settings and reputational data to be called from references and quality of scholarship.

University autonomy includes financial autonomy. Each university must generate sufficient resources to meet its expenses. According to the provisions of the Indian Constitution, the country has developed such a socio-economic pattern that Indian universities are not in a position to emulate the kind of autonomy that is found in countries like U.S.A. Here funding by the state/central government is essential. The governments may sanction block grants without strings attached. The universities will also have to generate new resources through a variety of programmes such as campaigning for donations, running short-term courses that are socially useful and valuable for various strata of society and professions, attracting research grants from industries, different ministries, commercial organizations and trusts.

University Teachers: Roles, Responsibilities and Service Conditions

Teachers in universities are appointed to teach, to conduct and supervise research keeping in view the objective of extending the boundaries of knowledge, to function as models and mentors of students, to observe and maintain discipline in the classroom as well as on the campus, to be available to students in their departments for consultations and counselling, and to evaluate students' performance towards the end of each academic session and assign grades in the courses taught by him/her. The role and responsibilities of teachers are composite and cannot be segmented and distributed, among teachers on extra-remuneration basis and still play the assigned roles effectively.

What is needed is to discontinue the practice of fractioning the roles and responsibilities of teachers by modifying the current concept of academic session, working time and service conditions. The first appointment as lecturer must not be in the university from where one has earned the qualifying degree, and it has to be a term appointment subject to extension for five years. If during this period one so appointed proves her/his worth, commitment, devotion to duty and potentials for research, he/she may be given tenure appointment. It is necessary to have flexibility about the nature of posts available in a department. Suppose a post of lowest cadre is vacant and the department is in search of a suitable candidate. In the process it is discovered that a highly qualified scholar is available deserving to occupy a Professor's chair. The executive council may allow such appointments on recommendation of the departmental council. In the same way against the post of professor a lecturer and reader may be appointed.

The concepts of academic session, holidays, suspension of classes, and examinations must be changed so that goals of academic excellence may be successfully pursued. The academic session for students should begin from 15th July and end on 30th April. For university personnel there should be no vacations except a few holidays such as three days of Holi, 3 days of Puja, one day each of Eid, and Moharram, Christmas, Independence and Republic days. Suspension of work on Jayanti days and visits of important political leaders is in no way justifiable. Every working day must be considered as a day for worship of the goddess of learning. The teachers should be allowed one month's earned leave and fourteen days of casual leave apart from medical leave.

A teacher's working day must begin at 9.00 A.M. and should officially end at 5.00 P.M. with lunch break for an hour. A teacher is required to be in the department during working hours. Apart from engaging her/his scheduled classes, he/she is expected to study, plan lectures, supervise research, be available to students for consultations and counselling and other academic work as planned by the department. Every teacher while on the campus is also required to see that work is going smoothly. Every teacher should be empowered to discipline students and keep students engaged in co-curricular activities planned by the department. Towards the end of the academic session a teacher is required to set question papers, hold examinations, evaluate answer-scripts and submit the award-lists to Chairman of the department in the case of master's degree students and to the Dean in the case of undergraduate students for publication of results. The entire examination processes, especially in this era of computers, must not take more than 15 days. The departments should have facilities and co-curricular programmes in the evening and educational trips on weekends for post-graduate students. Different departments, with the help of those members of the teaching staff who are not interested in taking earned leave during the summer vacation, must plan short-term courses for common people as an exercise in extension of the departmental expertise. Such activities hold promise of yielding, three pronged benefits to the university. The gown will prove useful in solving various problems of the surrounding communities, and brings them closer to the academia. Second, such programmes will add to university income. Third, university dons will get opportunities for sharpening their expertise and earn extra-remuneration also. Extra-curricular activities, such as games and sports, social service schemes, training programmes run by N.C.C. etc. are essential parts of university functions. For such activities it is not advisable to involve teachers for two important reasons. Primarily teachers are required to function as academicians. Assignment of extra-curricular work to them is not only inappropriate but also distractive and obstructive to teacher's involvement in curricular activities. Moreover, teachers are not appointed to teach on the basis of their expertise in the aforesaid extra-curricular activities. The most beneficial thing to do is to appoint full time experts in the various areas of required extra-curricular activities, each one being in-charge of a specific area under one of the faculty deans. These officers should be assigned duty and responsibility of (1) organizing and supervising games and sports, (2) functioning as the bridge between students and National Cadet Corps, (3) launching and getting social service programmes and getting them executed, and (4) searching out talented students and providing opportunities for sharpening their skills in various art forms. All these extra-curricular activities are to be conducted generally after classes are over on working days and on weekends.

Departmental Associations and Students' Unions

The traditional practice of constituting association of post-graduate students, one in each department, consisting of four or five elected office-bearers, is even today in operation, in spite of the fact that the academic standards and activities have fallen to abysmal depth of deterioration. These associations are created to organize a variety of co-curricular programmes under the supervision of a senior teacher of the department concerned. Such programmes include elocution contest, quiz, tests, creative writing competitions, festival celebration, welcome and farewell parties, collection of funds and academic resources for extraordinarily gifted but financially deficient students etc. Such associations are necessary part of department and to be further strengthened and provided with more facilities and incentives.

However, the current structure, function and dynamics

of university student unions are far from being conducive to the health of academic life on campuses. As a matter of fact they have become nurseries of the present day cunning and self-centred politicians for breeding new generation of political workers and troublemakers. The system of such unions must be abolished, and a new kind of student union must be conceived and developed. No one can deny the fact that a properly structured student union can serve useful purpose and supplement university education in building knowledgeable, skilled and hardworking citizens of a country.

In order to create a properly operative system of student unions in universities three stipulations must be made, *i.e.* (1) its membership should be restricted to undergraduate students only, (2) it functions under the guidance and supervision of a union officer to be appointed by the university on a tenure basis as one who has specialized in educational administration and social works, and (3) its activities must not be allowed to interfere in regular teaching work. Union activities must be carried out after teaching hours on working days and on weekends only.

Student unions should be constituted democratically through elections in accordance with the rules and procedures framed by university authorities, *i.e.* academic and executive councils. Office-bearers of the union, such as its president, vice-president, general secretary and faculty representatives may either be directly elected by all *bonafide* undergraduate students, or from amongst the elected representatives of different disciplines in various faculties of the university. After formal nomination of candidates, canvassing by them for garnering support of voters may be allowed to use slogan shouting, distribution of leaflets, visual displays and taking out processions, door to door canvassing with strict conditions that canvassing shall not be carried out during teaching hours and under no circumstances loud speakers be permissible. The office-bearers of the union should have the responsibility of planning as well organizing union activities and utilize budgetary provisions with prior approval of the union officer.

The main purpose of providing for formation of a student

union is to initiate students in democratic processes and methods used in solving problems of collective interest. It also created congenial conditions for actualizing their potentials for leadership, self-determination as well as self-regulation, keeping in view this broad objective a student union should be entrusted the responsibility of (1) publishing a weekly or monthly university news and view bulletin, (2) organizing cultural and charity shows, (3) running or looking after cafeterias, tea-stalls on the campus, (4) collecting funds for helping needy students, (5) looking after infrastructural facilities for the benefit of students, and (6) organizing seminars on problems of common interest so that the university dons and surrounding communities are brought together and interact constructively. However, it should never be allowed to become an instrument of political parties and to function as a pressure group for extracting anti-academic benefits. If and when the university administration and/or teaching community show apathy to students' genuine problems the union office bearer may submit memoranda or meet the official in groups, apprise them of their problems, and demand expeditious redress.

Meanings of Degrees and Division/Grades

In a variety of academic situations it is assumed that degrees, diplomas, divisions/grades, number of published papers and other certificates are reliable and valid measures of a candidate's academic merit because of the prevalent over-emphasis on objectified evaluation of scholastic performance in examinations that are conducted confidentially and anonymously. But the reality, so blatantly ignored, is that these measures vary in standards from one university to the other. So is the case with published papers. There are journals and journals, some of them publish research papers on payment by the author, and a few of them require papers so urgently that they publish any rubbish that is in the format of a research article. Therefore, these indices cannot, and must not, be considered as precise, reliable and valid measures of one's knowledge, ability and expertise. As a matter of fact, they are merely indicative of stages one has successfully crossed in various facets of educational development. They only qualify you for certain posts against which appointments are to be made on the basis of your degrees etc. and performance in the interview. Even this practice is now not considered useful in selecting really meritorious candidates.

So far as admission to higher degree courses, *i.e.*, master's and doctoral programmes is concerned, degrees and certificates from other universities should only make candidates eligible for seeking admission and not more than that. I am convinced that every department should hold a qualifying test focusing on candidates' comprehension and allow admission on the basis of performance in the test.

Concluding Remarks

Readers of this article may conclude that it is nothing but articulation of utopian thinking, and that the author has no contact with realities of Indian ethos. Suggestions made herein may be considered utterly impracticable by politicians in power and bureaucrats who have enjoyed pleasures of ruling over the university dons under the slogans of social justice and equal opportunity for all, may never allow university autonomy to be restored. A fairly large number of people may believe that no one should be fully trusted in matters of administration. Some may feel that the suggestions made here are too idealistic to be ever realized.

Undoubtedly cherished ideals are those that have not been realized. Once realized, all ideals become parts of social reality. At present, the concept of total autonomy to universities is worth cherishing and talking about. The more it is discussed there are greater and greater chances that it may become a topic for national debate, and some day something worthwhile may be done in this respect.

Index